F. Richards
F. H. Colvin

Practical Perspective

Elibron Classics
www.elibron.com

Elibron Classics series.

ISBN 0-543-70267-7 (paperback)
ISBN 0-543-70266-9 (hardcover)

This Elibron Classics Replica Edition is an unabridged facsimile
of the edition published in 1908 by The Norman W. Henley Publishing Co.,
New York.

Practical Perspective

A TREATISE SHOWING JUST HOW TO MAKE ALL KINDS OF MECHANICAL DRAWINGS IN THE ONLY PRACTICAL PERSPECTIVE (ISOMETRIC). MAKES EVERYTHING PLAIN SO THAT ANY MECHANIC CAN UNDERSTAND A SKETCH OR DRAWING IN THIS WAY. SAVES TIME IN THE DRAWING ROOM AND MISTAKES IN THE SHOPS. CONTAINS PRACTICAL EXAMPLES OF VARIOUS CLASSES OF WORK.

BY FRANK RICHARDS

Associate Editor American Machinist

AND FRED. H. COLVIN

Third Edition *Fully Illustrated*

NEW YORK:

THE NORMAN W. HENLEY PUBLISHING CO.

132 NASSAU STREET

1908

The Principles of Isometric Perspective

BY FRANK RICHARDS
Associate Editor "American Machinist"

▲ ▲ ▲
▲ ▲

I suppose it is from constantly seeing working drawings of machinery, and no other representations of it, that mechanics generally are entirely unable to sketch anything in perspective. Yet, in perspective or pictorial work is where free-hand drawing is most valuable to the mechanic. Without perspective or without some representations of more than one side of an object, a drawing is not readily and effectively descriptive to the untrained eye. It seems to me that if mechanics could get hold of the simple principles of isometric projection, it would help them much in making intelligible sketches of machinery. The mechanical draftsman does not use it very commonly, and I suppose the principal reason is the difficulty of drawing ellipses. But this objection does not apply to free-hand drawing at all, because it is just as easy to draw an ellipse by hand as it is a circle, and perhaps a little easier. So I will venture to explain the principles of isometric projection, so that common folks will not be afraid to use it every day, if they choose to.

The regulation way of teaching the principles of isometric projection is by the drawing of a cube (Fig. 1).

That is supposed to show the whole thing. Of course it does; but why not use a brick, that shows it better? Even the word cube is objectionable, because it is one of the words that frighten the poor little machinist. But, seriously, a brick is preferable for our purpose, because its

Fig. 1 Fig. 2 Fig. 3

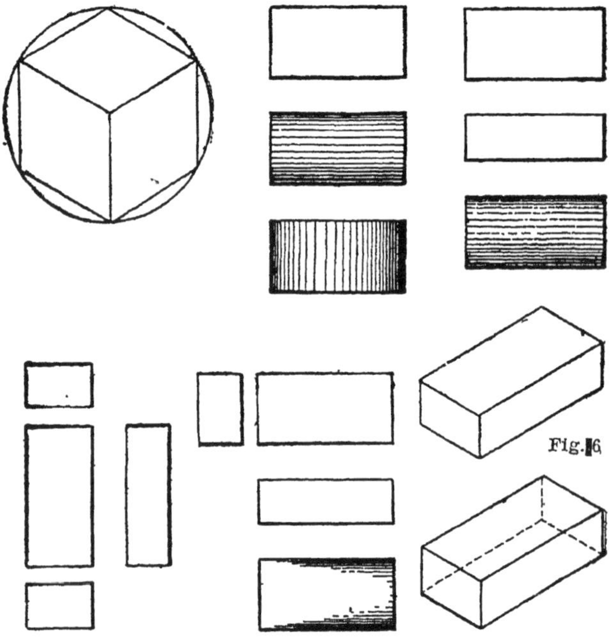

Fig. 6

Fig. 5 Fig. 4 Fig. 7

three dimensions are all different, and cannot be mistaken for each other.

Now, if we were going to make a working drawing of a brick, we couldn't get along without at least three views of it: plan, side or front elevation and end elevation. Without all three views we would be very much in doubt about it, and even with all three it would be possible to be mistaken; so one is really not very much to blame, nor very much to be despised, if he is not quick to see the thing that a "draft" tries to represent. With only the plan before us (Fig. 2,) the thing represented might be cylindrical or round in either direction. With plan and side elevation (Fig. 3) it might be elliptical in cross-section. With all three views (Fig. 4) it might have the straight lines and the square corners at the end shown and the corners might taper off and the other end of the object might be oval or semi-circular, and the drawing would still be a correct representation of it. So to bind the meaning of the drawing beyond all possibility of mistake, we would require a view of the other end of the brick (Fig. 5) or four views in all. But with an isometric projection of the brick (Fig. 6), we would require no label to tell us, "This is a brick," and we wouldn't find it possible to mistake it for anything else. By drawing the dotted lines (Fig. 7), the entire outline of the brick is shown. This is another advantage the isometric projection often gives us.

It will be remembered that I am calling attention to isometrical projection for its value and applicability to sketching or free-hand drawing. The little illustrations that I offer are made with instruments in the regular mechanical way, and that may be said to be rather inconsistent. But I am only trying to put *the idea* of isometrical drawing in a simple and familiar way, so that common, every-day, young shop folks will be able to get hold of it. The application of it will come readily enough to any one who once fully understands it; and it will come in play whether making working drawings with instruments or sketching with eye and hand unassisted.

We are now ready to make an isometric projection of a brick. Our brick, we need not remind any one, has three dimensions: length, width or breadth, and thickness or depth. Either dimension may be assumed for the vertical one. We can draw the brick as standing upon its side (Fig. 8), its edge (Fig. 9), or its end (Fig. 10); and so the brick can be drawn in isometrical projection in three different positions, or, making them right and left, we can have six different views, using whichever may be most convenient. Strictly speaking, our brick can be drawn in any position, as, for instance, where it might occur in the setting of a boiler; but it will be better for us not to say anything about that now.

We can draw our brick first in what we may call its

Fig. 8

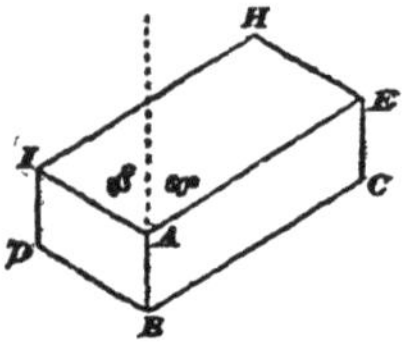

Fig. 10

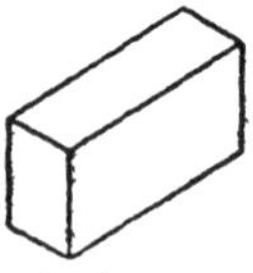

Fig 9

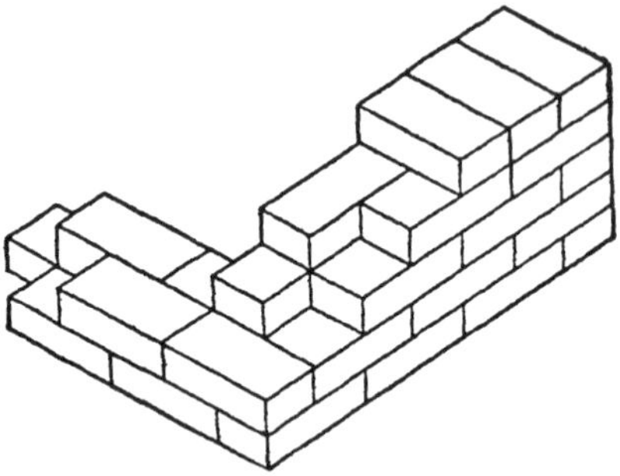

Fig11

most natural position—lying upon its side. Then the lines running in the direction of its thickness or height, the corner lines, will be vertical. Lines that are vertical in the object to be drawn should always be vertical in the drawing. This law determines the direction of many lines of a drawing, without any further trouble, and it may usually be well to draw a prominent vertical line of an object for a beginning. The character of the object will decide that. Having determined the scale that we want to make our drawing to, we can draw the first vertical line *A B*, according to that scale. There is a scale of some kind to every drawing. When we set out to make a rough sketch of anything, we have some idea of how big we want the whole thing to be, and then, from the beginning, we make the parts of it of a size or length to correspond with our idea of the whole, and that is working to a scale in a rough and perhaps unconscious way.

Now when we come to draw our next line we come to one of the peculiarities of isometrical drawing, and that is that straight horizontal lines in the object that stand square with, or at right angles to each other, are drawn in one or the other of two fixed directions. We draw the line *B C* for the bottom corner of the front edge of the brick. This line should stand at the angle of 60 degrees from the vertical line, and all lines in the brick running lengthwise of it, and that are parallel with the one represented in the drawing by our line *B C,* must be drawn parallel with that line, or they must all stand at 60 degrees from the vertical line. Then the line *B D* for the bottom corner of the end of the brick must be drawn in a direction 60 degrees the other side of the vertical line. In following the isometric system for free-hand drawing, it of course will not be necessary to determine either of these angles with any accuracy.

We have now the three lines *A B, B C,* and *B D. A B* is already of the required length according to our scale. We now measure off the other two lines to the required length according to the same scale, and that brings us to the other peculiarity of isometric drawing, the peculiarity that is supposed to be described in its Greek and Latin name, the peculiarity that makes it peculiarly valuable for mechanical drawing from true perspective, all the rectangular lines, all the lines of a brick, in an isometric drawing are of their proper proportional length, according to the scale of the drawing. An isometric drawing is thus a correct, accurate and reliable working drawing, which a true perspective drawing can never be.

To complete our drawing of a brick no further measuring will be required. From the point *A* we draw *A E* parallel to *B C*, and *A I* parallel to *B D*; then the two vertical lines *I D* and *E C*; then *E H* parallel to *B D* and *I H* parallel to *B C* finishes the work.

That is all there is of it, and there is nothing terrible about it. Anybody can project a brick. And anybody that can project a brick on paper can draw any rectilinear rectangular figure. And now, you see, I am in trouble again. He will surely take fright at those two terrible words. I mean that, as far as we have got now, you can draw anything that is made of straight lines that are parallel to or square with each other. We can now make an isometric projection of anything that is made of parallel lines and right angles. Where oblique lines and circles come into the case we will have a little more trouble, and we will not say anything about it now. We will exercise what we have so as to fasten it. If you can draw one brick you can draw several, and I have piled up a few here (Fig. 11) that ought to make a good exercise. You can find plenty of real models around the shop, and they will be of use to you if you use them.

It will be remembered that I started about this isometric drawing as useful in the free-hand drawing of machinery; yet I am making so much fuss about it that I am afraid I will frighten the sketcher away. But while we are on any subject it is always proper to say that the more we learn about it, and the more thoroughly we understand it, the more useful it will be to us.

The trouble about the isometric business begins when

we go to represent anything other than the rectangular lines. It is just like it is in life, you know. Anybody that is not square is bound to learn what trouble is. Now, suppose that we have two bricks, (Fig. 12), one lying flat and the other one tipped over against the end of the first

Fig. 16

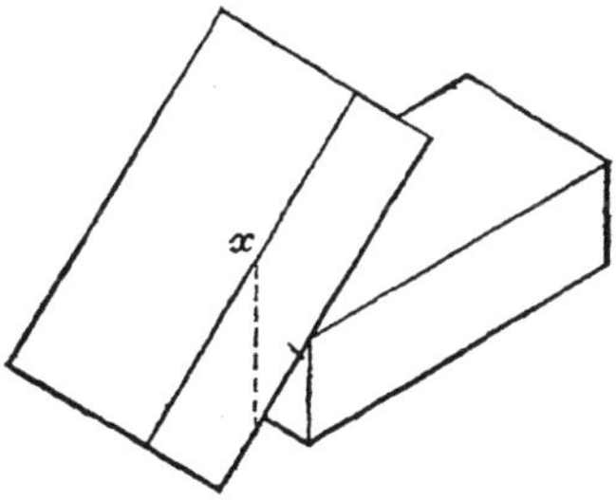

Fig. 12

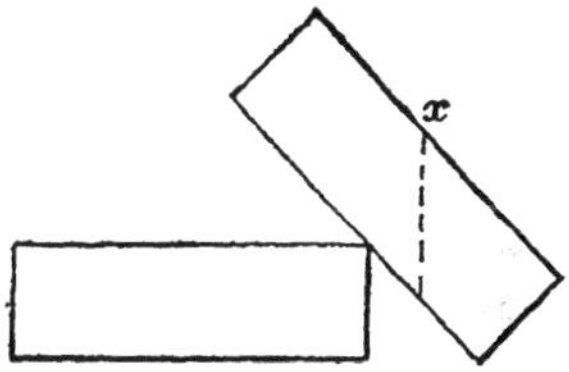

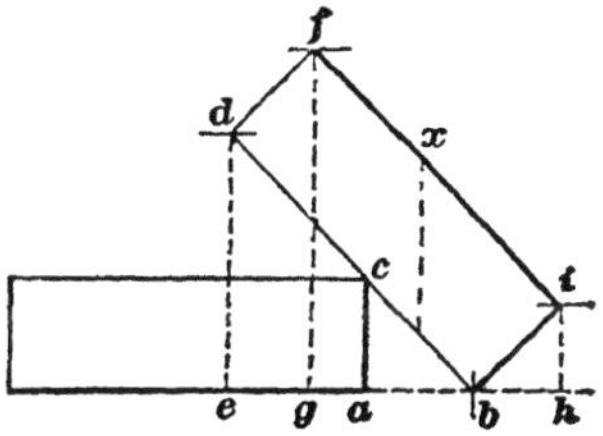

Fig. 13

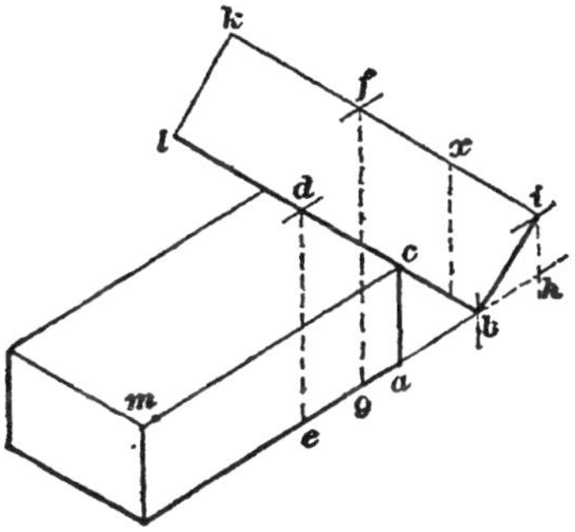

Fig. 14

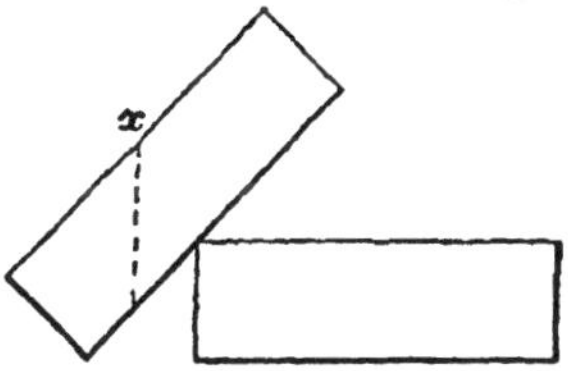

Fig. 15

at an angle of 45 degrees, the front edges of both bricks being in the same plane; that is, if you laid a straightedge against the front edge of both bricks, and in more than one direction, it would always touch and coincide with both. (This is a complete scientific definition, just disguised enough for me to get it into anybody without frightening them.) Now, none of the lines defining the outline of the front edge of the inclined brick, nor those at the back edge parallel to them, will be isometric lines, and we must find some way of getting at both the length and direction of them. The remaining lines of the end will be isometric lines, and will be parallel to the end lines of the other brick. When we get the location of the angles from which these end lines start, we can draw them to the correct length, according to the scale of the drawing. All these drawings of the brick I make to scale, and I call the dimensions of the brick 8″x4″x2½″, and the scale is ⅛. In Fig. 13 we prolong the base line of the under brick as far as seems necessary, and we make the isometric drawing of the under brick, Fig. 14, and prolong the base line of that, also. Then in Fig 13 we measure off the distance, *a b*, and mark the point, *b*, in Fig. 14. As the angle in this case happens to be 45°, and as the height of the under brick is 2½″, we know that *a b* will be 2½″. Then a line drawn from *b* and just touching *c*, will be the first line of our other brick; but we don't know the length of it. In Fig. 13 we let fall the perpendicular *d e*, which gives us the distance *e a*. Then measuring the distance *e a* in Fig. 14, we raise a perpendicular to the height *e d* obtained from Fig. 13. Then *d* will give us the length of *d b*, which we need not have drawn until now. In the same manner we may obtain either the point *f* or the point *i*, say *d f*; then the other two sides *f i* and *i b* may be drawn parallel to *d b* and *d f*, and we have the edge complete. *f k* and *d l* being isometric lines, we may draw them at the required angle, and to the length according to our scale, and then drawing *k l* will finish the whole job.

A peculiar property of the isometric projection is brought out by the angle we in this case have assumed for the inclination of the brick. The angle being 45°, but two of the six faces of the brick appear. This could only occur at this precise angle. If the brick were tipped down more, so that the angle with the base line became more acute, the upper face of the brick would begin to show, and if the brick were tipped up so that the base angle became greater, we would begin to see the under face of the brick.

Another peculiarity that the use of the angle 45° brings to our attention, is the relation of the line *f i* to *k f*, and of *d b* to *l d*. That they should be in a straight line is, indeed, involved in what we were thinking of in the last paragraph—the fact that at this angle of inclination the upper and under faces of the brick are both invisible, and those faces being planes, and coincident with the lines extending to the eye of the observer, the entire visible boundary of either face must be one straight line. You take anything with a perfectly flat surface, no matter what the shape of its outline, round, square, triangular or irregular, and tip it up until the face just disappears from sight, and all the edge of that surface that you can see will be a straight line. But the queer thing about our brick is, that one part of the straight line is an isometric line, and another part of the same line is not. *f k* and *d l* are true isometric lines. Their direction is correct, and their length agrees with the scale of the drawing. But *f i* and *d b* are not isometric lines. Any one can see how they are for length, without taking the trouble to measure anything. *d b* should be of the same length as *c m*, but it is evidently shorter than that. So, too, *i b* should be of the same length as *c a*, but any one can see that it is longer.

To show what queer capers the isometric drawing will cut with an object, I have taken the same brick and slid it along and tipped it up against the near end of the under brick, instead of the far end, and at the same angle, 45°, as before. It is drawn correctly according to the same scale as Fig 14. The distortions of dimensions, it will be noticed, are here reversed. The brick that was too short in Fig. 14, is now too long, and where it was too thick in Fig. 14, it is now too thin. In an isometric drawing, then, to have a line true to scale, it is not enough that it should be parallel to one of the three isometric lines. We must also know that the line represented stands in the object in one of the three rectangular directions that the isometric lines represent. We know that the front outlines of our leaning bricks were, at 45°, equidistant from the two, and, consequently, as wrong as they could be in that plane. It is the same with imaginary lines as with actual lines. Let me call attention to the vertical dotted line marked *x*. It is of the same length in all the figures, and in the isometric projections it is a true isometric line, and is correct to scale in length, except that in the thin brick, Fig. 16, it is apparently a little longer than in the thick brick, Fig. 14; but that is only because I am not the most accurate draftsman in the world.

It is proper, also, to call attention to the angles in isometric drawing. As a general rule, we may say that they are always wrong. They are not, in the isometric drawing, the same as in the object represented. We have been representing our brick tipped over at 45°; but that is not the angle in the drawing. And, curiously enough, the corners of the brick itself, that is tipped over at 45°, are correct. The brick being wrong, its angles are right. But in the under brick, which we assume to lie on level ground, none of its angles are right angles. If we ever become expert enough in isometric drawing to dare to tackle circular work, we may learn more of how angles are squashed and jammed out of shape by isometric projection.

From all these things we may infer that an isometric drawing is fearfully and wonderfully made. It is not a thing to be trifled with. As a working drawing in a shop it wouldn't do for an ignoramus to try to measure it anywhere. For the working measures of parts he must follow implicitly those that have been written in by competent and responsible persons, and stray not either to the right or to the left. I have no expectation that isometric draw-

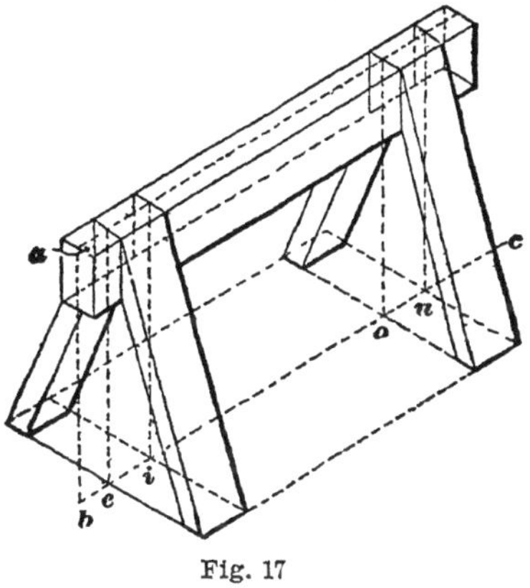

Fig. 17

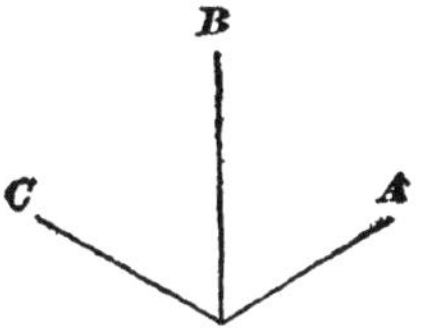

Fig. 18

ing will be generally adopted for working drawings, and I do not advocate it; but I do know that it would enable many a workman to get an idea of the shape of a piece, and of how the different pieces of a machine are related. Like everything else, it is valuable in its place, and would be more generally used if mechanics, and especially young mechanics, would take the trouble to get the hang of it.

Generally, in drawing machines or parts of machines in isometric projection, no plane projection, for the purpose of getting construction lines, or of locating points, will be needed, as there will be rectangular lines in the thing represented, with which any oblique lines that occur will be in contact, or to which they can be referred.

Here is a horse, a familiar quadruped of the shop, (Fig. 17). To draw it nothing is needed but the single projection and such construction lines as we can readily locate by it. We need the measures of the parts, of course, but those we need in any case. The body of the horse is rectangular, and we draw the entire outline of it according to our scale, the parts out of sight being in pencil in our drawing, and shown here by dotted lines.

It occurs to me, just here, that we ought to have a name, as a labor-saver, by which to designate the three normal isometric lines, *A, B, C,* Fig 18. The vertical is, of course, self-named. The other two we may call the right isometric line, *A,* and the left isometric line, *C;* and so I use them.

Well, from the center of the near end of the body of the horse we drop the vertical line *a b,* the length being equal to the height of the horse. Then from *b* we draw the right isometric line *b c,* which will be the center line of the base. Along this center line we measure and mark the four points *e, i, o, u,* corresponding with the same points upon the top of the horse. Through each of these we draw a left isometric line, and space one of them equidistantly upon both sides of the center line, the sum of the two measures being equal to the spread of the legs at the base. Then by two right isometric lines we carry these same measures of the first left isometric line to the other three of them. The vertical lines at the sides of the body where the legs meet it are then to be indicated. We have now all the marks necessary to complete the horse, which we proceed to do by drawing the remaining lines. The outer corner lines of the legs being drawn, the inner lines are simply drawn parallel to them, and meeting the body at the under corner of it.

The phenomenon regarding the change of oblique dimensions is again here to be observed. The length of the front legs is too long, and the length of the back legs is too short. This is not at all due to the distance that they stand from the eye, as would be the case in true perspective, but is entirely on account of the different angles at which the legs stand. There seems to be a law of compensation in the matter, too, for, where the leg is elongated

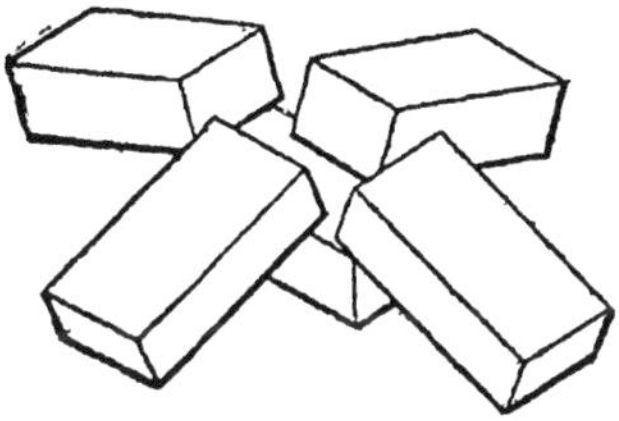

Fig. 19

Fig. 20

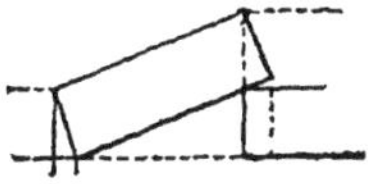

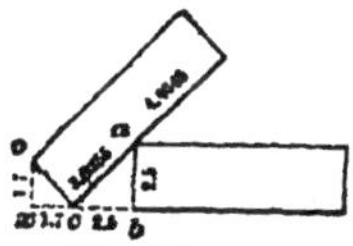

Fig. 21

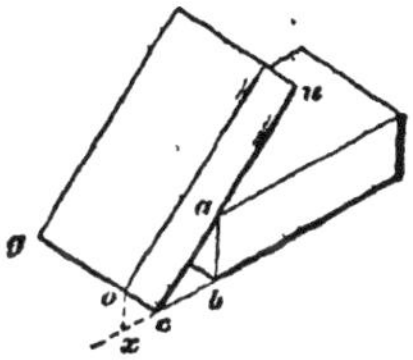

Fig. 22

it also becomes thinner, and where it is shortened it becomes thicker.

Here are four bricks, Fig. 19, placed centrally and rectangularly, and at the same angle of inclination upon the two edges and the two ends of a brick lying flat upon its side. This is an easy thing to draw, the only assistance required being one elevation, Fig. 20, of the reclining brick, from which all necessary measures may be taken. All four bricks reclining at the same angle, the bottom corner of each is at the same distance from the under brick, and the upper corners are all of the same height. Fig. 19 shows rather strikingly, I think, how the isometric projection may change the appearance of an object. The bricks at the back would scarcely be accepted by the uninitiated as identical in size and position with those in front.

In practical drawing, the same as in machine work, in the preliminary construction of a machine upon paper, as in the actual work of the shop, the faculty of invention is always in play. Different ways of doing things are always to be thought of, for it is seldom that the best is thought of first. As I believe in wringing things as dry as possible, as I always advocate the getting as much as possible out of everything before leaving it, I may be permitted to follow the practice that I advocate, and bring up what we have already mentioned before. In the drawing of the brick, (Fig. 16), inclined at 45°, instead of drawing any of the dotted construction lines, as we did before, except the base line, we could have got all the required points by arithmetic; by the arithmetic of the common school that the mechanic should carry into the shop with him when he enters it. I always say, as I always believe, that arithmetic is as valuable to the mechanic as any tool he can have, and, like all his tools, it is not of much use to him, and does not speak well of him if it is allowed to get rusty. So it will be good for it and for us if we brighten it up a little by exercising it in this case. This will give us quite a neat little job in arithmetic, bringing into play the three most important arithmetical operations the mechanic has to use: the extraction of the square root, finding the third side of a rightangled triangle, and simple proportion. I would like to say more about machine shop arithmetic by and by, if we travel together so far, but just now we will stick to the brick case.

The angle of inclination, Fig. 21, being 45°, *b c* must be equal to *a b*, and in Fig. 22 we can measure off *b c* and then draw *c a*, and prolong it upwards beyond *a*, as far as may seem necessary. This will be the lower line of the inclined brick, but not being an isometric line we cannot measure off the length of the brick by our scale. We first find the length of *a c* in Fig. 21. This is the hypothenuse of the right-angled triangle *a b c*, so we get $2.5^2+2.5^2=12.50$, and $\sqrt{12.50}=3.5355$, the length of *a c*. Then the length of the remainder of the brick will be $8—3.5355=4.4645$. Now this, instead of being the end, is only the beginning. The length *a c* in Fig. 21 is correct according to our scale, but in Fig. 22, while *a c* represents the same portion of the total length of the brick as *a c* in Fig. 21, its actual length is not the same. It is evidently longer, and we want to find what its actual length is. To show this comfortably without hitting anything, we draw this on a larger scale, Fig. 23. Now, from *c* as a center, with the radius *c b*, we draw the arc *b i d* and the line *c d*, making the angle *i c d* 30°, the same as we know that the angle *i c b* is. If we have been studying our geometry, as I assume that we have been, we know that *b d* is equal to *c b*, and consequently *e b* is equal to one-half of *c b*, or 1.25. Then from the right-angled triangle *b e c* we get $2.5^2—1.25^2=4.6875$ and $\sqrt{4.6875}=2.16$, the length of *e c*, and from the right-angled triangle *a e c* we get $(2.5+1.25)^2+2.16^2=18.7281$, and $\sqrt{18.7281}=4.32$, the actual length of *a c*. But this actual length (by the scale of the drawing) of *a c* represents only 3.5355, and if 4.32 represents 3.5355, what will represent 8, the total length of the brick? This we find by the simple proportion: 3.5355 :8 :: 4.32 :? and solving this we have $8\times4.32\div3.5355=9.77$. This 977 then will be the length *c a u* in Fig. 22.

To complete the brick we have now to locate the point *o*, after which the remainder of the figure is determined by lines drawn parallel to those already made. In Fig 21 *o x c* being a right-angled triangle, and the angle *o c x* being, as we know, 45°, and the side *o c* opposite the right angle being 2.5, we get $\sqrt{\frac{2.5^2}{2}} = 1.7$, which will be the length of both *o x* and *c x*. These lines will be, in Fig. 22 or 23, true isometric lines, and drawing them according

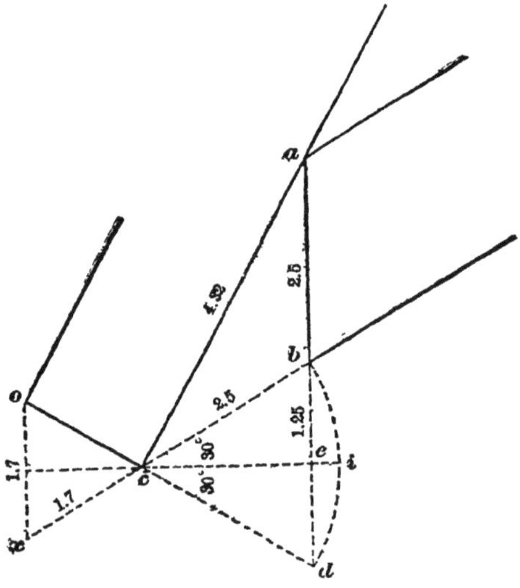

Fig. 23

to our scale gives us the point *o*. From this we draw *o g* to scale, and then the parallel lines that finish the brick.

We may now venture to look a little at the standard and traditional diagram that has always been used to show the principles of isometric projection. It is very convenient for the purpose, and I have not meant to treat it with any disrespect. Here is a plane projection of a cube, Fig. 24. Of course all its sides are equal, and all its angles are right angles. Here is the isometric projection of a cube, drawn to the same scale, Fig. 25. The sides are still equal to those in the plane projection, and still equal to each other; but the angles are changed. Two opposite angles of each face of the cube have become obtuse angles, and the two alternate opposite angles have become acute angles. The square has become a rhombus.

Now no mechanic has any respect for a rhombus. He always thinks of it as a square that has been spoiled in making. It has tipped over, (Fig. 26), before its joints have set, you know. But the same mechanic has great admira-

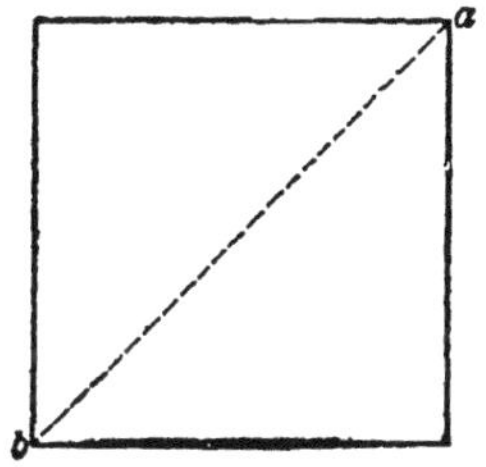

Fig. 24

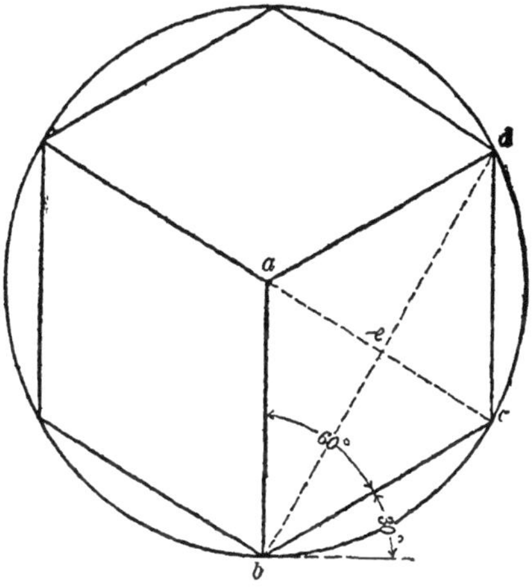

Fig. 25

tion for the diamond; so here we present him with a couple of them, Figs. 27 and 28. You will have hard work to persuade him that they are the same old rhombus in disguise. It may be that the disguise was the other way, and that what we habitually think of as the diamond is the real thing. We certainly know more of its properties under that name. We regard it as one of the most symmetrical of figures. It is symmetrical in relation to both of its axes or center lines. When a square becomes a rhombus, as it does when we make an isometric projection of a square surface, the thing that occurs in connection with the change of the angles is a change in the length of the diagonals, or the lines connecting the opposite corners of the square. One becomes longer and the other becomes shorter. But, however much the square may be tipped over, and however much the angles may be changed, the two diagonals always remain perpendicular to each other. We encounter this principle frequently in machinery, in "lazy tongs" and movements of that character. This was why the corners of our brick, tipped over at 45°, appeared as right angles. The sides and ends of the brick had then come into the position of the diagonals of a square.

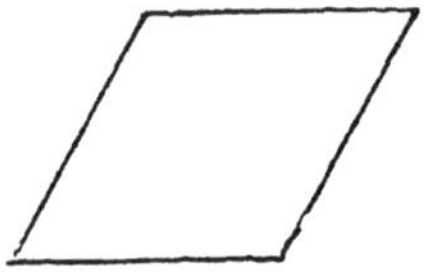

Fig. 26

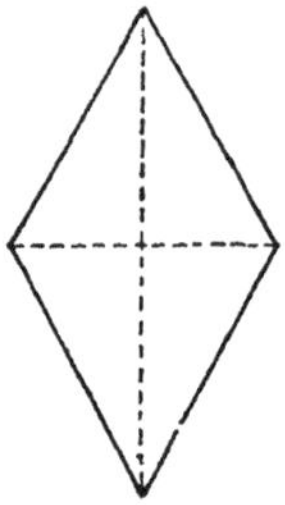

Fig. 27

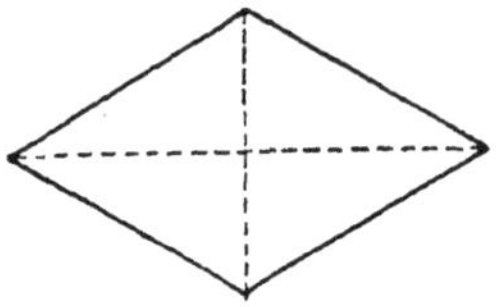

Fig. 28

The rhombus that occurs in true isometric projection always has the same angles, so we need not chase our diamond to any extremes. In isometric drawing the proportional length of either diagonal of the rhombus to the side of it is always the same, and the proportional measure of any figure, when measured in a direction parallel to the diagonal, will be changed in the same ratio. A circle lying in the plane of either square face will be elongated in the line of the longer diagonal, and in the same proportion as the diagonal is itself elongated, and it will be shortened in the line of the shorter diagonal. The circle becomes an ellipse, and it is an ellipse of constant shape. If we can find the constant relative lengths of the two diagonals we can know the constant ratio of the two axes of the ellipse to each other, and the ratio of each to the actual diameter of the circle represented.

The diagonal *a b* of the square, Fig. 24, being the hypothenuse of a right-angled triangle, and the other two sides each being 1, the diagonal will be $\sqrt{2}$, or 1.414. In the isometric drawing of the square, the right isometric or base line *b c* being 30° from the horizontal, and *a b* being vertical, the angle *a b c* must be 60°. Now we know, because everybody knows, or ought to, that 60° is the angle of an equilateral triangle, and the triangle *a b c* must be such a triangle. Then *a c* must be the same length as the other sides of the triangle. Its length must be 1. But its original length in the square was 1.414, and its present length is a fraction of its original length represented by $\frac{1}{1.414}$, and the measure of everything on this line will be this fraction of its original length. To obtain the length of the long diagonal we have here another right-angled triangle *a d c*, that will readily give us half of it, or the same triangle would do it. The length of *d c* is 1, and *e c* of course .5, then the length of *d e* is $\sqrt{1^2 - .5^2} = .866$. This being one-half the long diagonal, its whole length must be 1.732. Its present relative length, then, treated the same as we served the short one, will be $\frac{1.732}{1.414}$. If now we make both of these into decimals, by dividing the numerator of each by its denominator, we get .707 and 1.224 as the proportional (not actual) lengths of the diagonals, and the proportional lengths of the axes of the ellipse will be the same. These will also be the actual lengths of the axes where the original diameter was 1.

As I believe I said in the beginning, the difficulty of readily describing ellipses has had much to do with preventing the general use of isometric projection. If an ellipse could be described as readily as the circle can be drawn with compasses, we would use it more, and we would know more about it. I remember that I learned much of the properties of circles by scratching them on the ground with the tines of a pitchfork. An ellipsograph is too expensive a plaything for every boy to have; and the best of them, so far as I know, have the fatal defect of not making a good line in ink. They will describe an ellipse very nicely with a pencil, but, after all, the ellipse must be inked by hand, so that what we use to ultimately guide the pen in inking, we use for the whole process, and we put the ellipsograph carefully away upon the top shelf.

The ellipses made in sets that I find in catalogues of drafting instruments are not of the right proportions for isometric drawing. In those that I have examined—and I know of no others—the short diameter is .75 of the long diameter. The short diameter, the long diameter being 1, of the isometric ellipse, is: $\frac{.707}{1.224} = .579$. The scale which I offer, Fig. 29, half size, should be of some use. This will readily show the semi-diameters of any ellipse within the range of it. The base line represents the actual diameter of the circle, and upon it may be raised vertical lines representing as many subdivisions of the inch as may be desired. Where these verticals are cut by the two oblique lines will be the corresponding semi-diameters of the ellipse.

The ellipse, Fig, 30, will touch the sides of the rhombus, as the circle touches the sides of the square, at the middle of each. Then if upon the two diagonals we measure off the extreme dimensions of the ellipse, as obtained

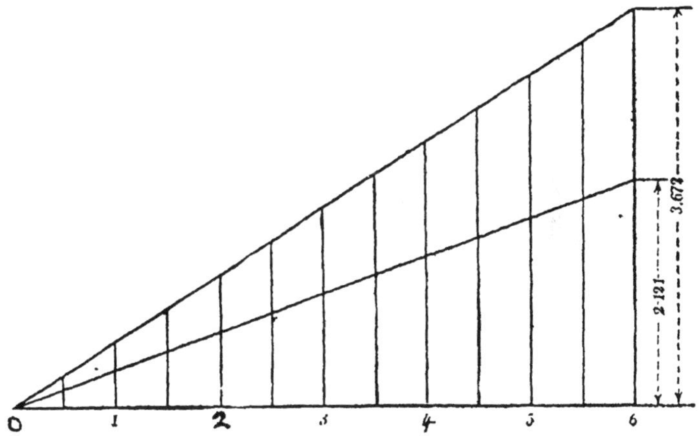

Fig. 29

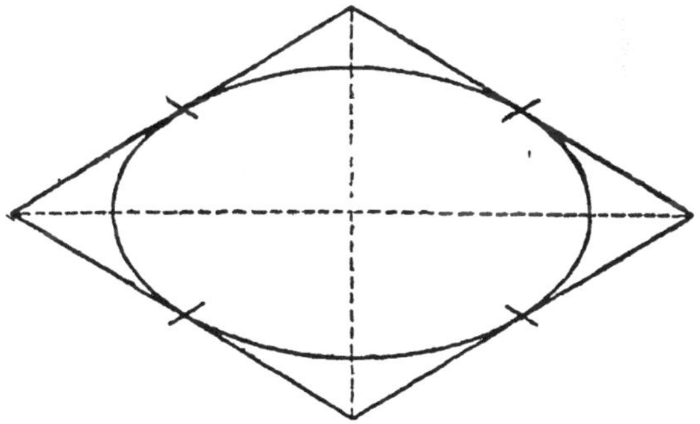

Fig. 30

from the scale, Fig. 29, or otherwise, we have eight points of the ellipse, and we can readily draw it in with sweeps.

If eight points are not thought to be enough to define the ellipse accurately, we can readily obtain as many as we choose. One method would be by the subdivision of the square into any number of smaller squares, and a corresponding sub-division of the rhombus. This would apply to the reproduction of any figure, regular or irregular, in any isometric plane. The five-pointed star, Figs. 31 and

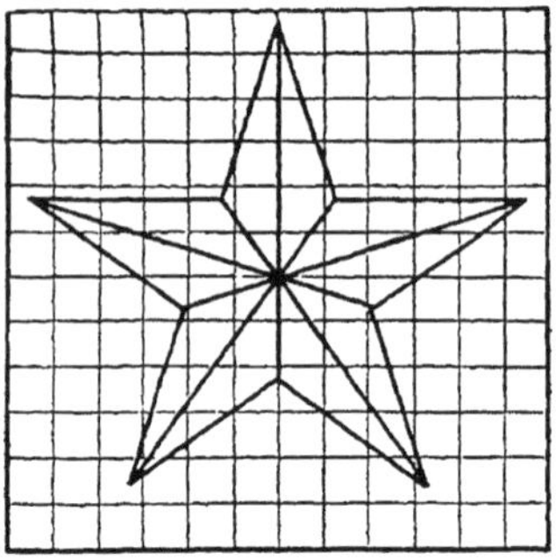

Fig. 31

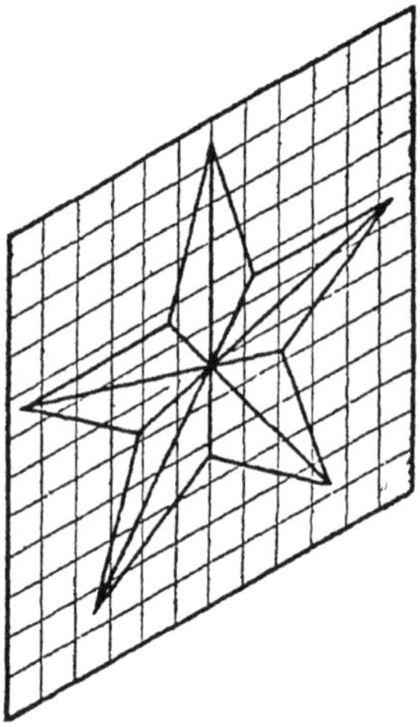

Fig. 32

32, is not difficult, because only the five extreme points have to be located, the straight lines determining all the rest, and the system of squares was not necessary. By the way, Fig. 30, by turning it around, will correctly represent either face of the cube.

The Use of Isometric Paper

BY FRED H. COLVIN

Mr. Richards has explained the principles of isometric perspective, which is the only practical perspective for the mechanic, and has shown some of its applications to machine work. That it has not become much more common can only be due to the fact that it has not been understood, that it was too much trouble to get at it and the trouble of laying it out on ordinary paper.

By the use of the D-C Isometric Sketching Paper nine-tenths of the difficulties disappear and a sketch can be made in short order without the use of anything but a pencil as even a rule or a 30 degree angle is not needed. If a finished drawing is desired you only need a compass and a rule in addition to the ruling pen as the sectioning of the paper takes care of all the rest.

After what has gone before it will only be necessary to give a few suggestions and show their actual application by examples. Free hand sketches are perhaps the most useful in nearly all cases and can be made readily, but if correctness and finished drawings are desired they can easily be obtained.

One of the first things to get straight in your mind is the proper position of the ellipses which represent the circles. Unless this is done some queer errors occur, but a few moments will make it so clear that you will never make a mistake in the matter. To show this clearly we will take up the old familiar cube which becomes a hexagon in isometric perspective as shown in Fig. 33, and we can make this clear.

Each side of the cube becomes a diamond as will be seen, and the ruling of the paper which is shown to make things clear, shows you exactly how these diamonds are located in each case. Each side of the cube is nine spaces long, and remembering that the dimensions are only to be taken along the isometric lines we have no trouble in laying off all three sides of the cube or anything else. The ruling forms these diamonds in the three directions required for showing square or round surfaces in isometric perspective as we shall see in every case. These diamonds are drawn a little more lightly than the ellipses which it is especially intended to illustrate.

As each side of the cube is nine spaces long the diameter along the isometric lines at *A* to *B* in all three cases will give the true diameter. This gives us four points in the ellipses, or in other words the center of each side of the diamond gives a point. To obtain the other four points, the long and short diameters there are several methods, two of which are shown here.

A circle struck from *O* as a center and touching the sides of the hexagon gives very nearly the outer curve of the ellipse in all three diamonds and also locates the outer side of the short diameter. The same radius from the outer corner will give the other side.

The top ellipse shows one way of finding the other diameter, which is easy but not quite correct enough for use if you want to lay out work from it. This is to draw a line from the middle of the side of the diamond or from the point already found, and use the point where this

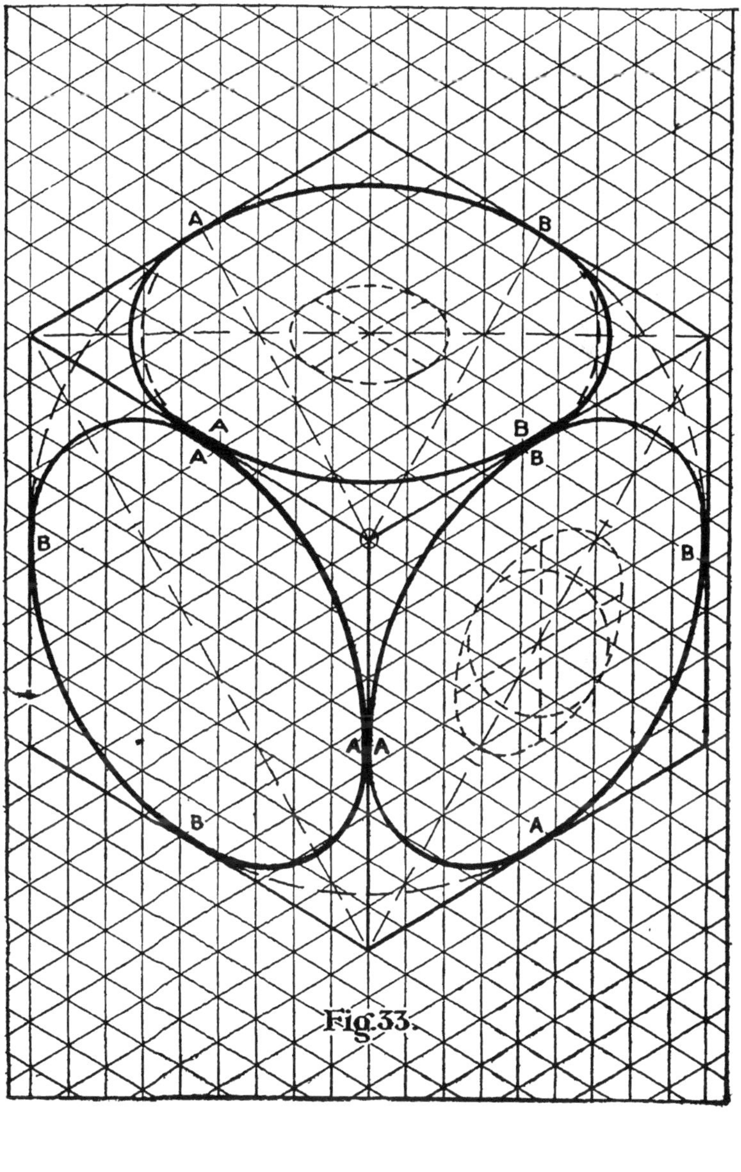

Fig. 33.

crosses the center line as a center for the arc. This gives too large a curve and too short an ellipse, as shown by the dotted lines at the end.

A much better way is to draw a circle representing the small diameter, as shown at the right, and where this cuts the center line is the center for the end curve. This construction, which is better in every way, involves the use of the diagram in Fig. 34, which explains itself and which can be used more easily than any set of curves. For most work you do not need extreme accuracy and you soon get to know the right proportion by noting the diamonds around or in which it is drawn; for it will usually come right to use them. The diagram just referred to also gives the correct radius for the larger and smaller curves and will be a great time saver where accuracy is required. The correct outer curve is not exactly as appears in this figure but very near to it. The diagram, (Fig. 34), is not claimed to be new with the exception of giving proper radius for each curve.

There is one vital point to be remembered regarding these ellipses and this is which to use to represent the view you desire, but this is easily learned. Whenever it is a top view or shows a circle on the top of any flat surface, the ellipse is *always* as shown on top of the cube or in the horizontal diamond. Don't try to twist it around because the rest of the piece is off at some angle or other for it belongs just where it is shown and will look all right when you get it done. If the rod or shaft or pipe runs up and to to the left, the ellipse, showing the end of said rod, shaft or pipe, will always be in the right hand diamond and the opposite is true if it runs up and to the right. If it runs down this is reversed. Remember this and you'll have no trouble on this score. On the paper itself the

The Use of Isometric Paper.

A—Diameter of Circle or Side of enclosing Diamond.
A to B—Short Semi-Diameter of Ellipse.
A. to C—Long Semi-Diameter of Ellipse.
A to D—Long Radius for Drawing Sides of Ellipse.
B to C—Short Radius for Drawing Ends of Ellipse.
a b c—half ellipse for 2-inch circle.
d e f— " " " 3 " "
g h i— " " " 4 " "
2 j—true radius of 2-inch circle.
3 k— " " " 3 " "
4 l— " " " 4 " "

These are all 30 degrees from long diameter. True dimensions are always along isometric lines; 30, 60 or 90 degrees from horizon or vertical.

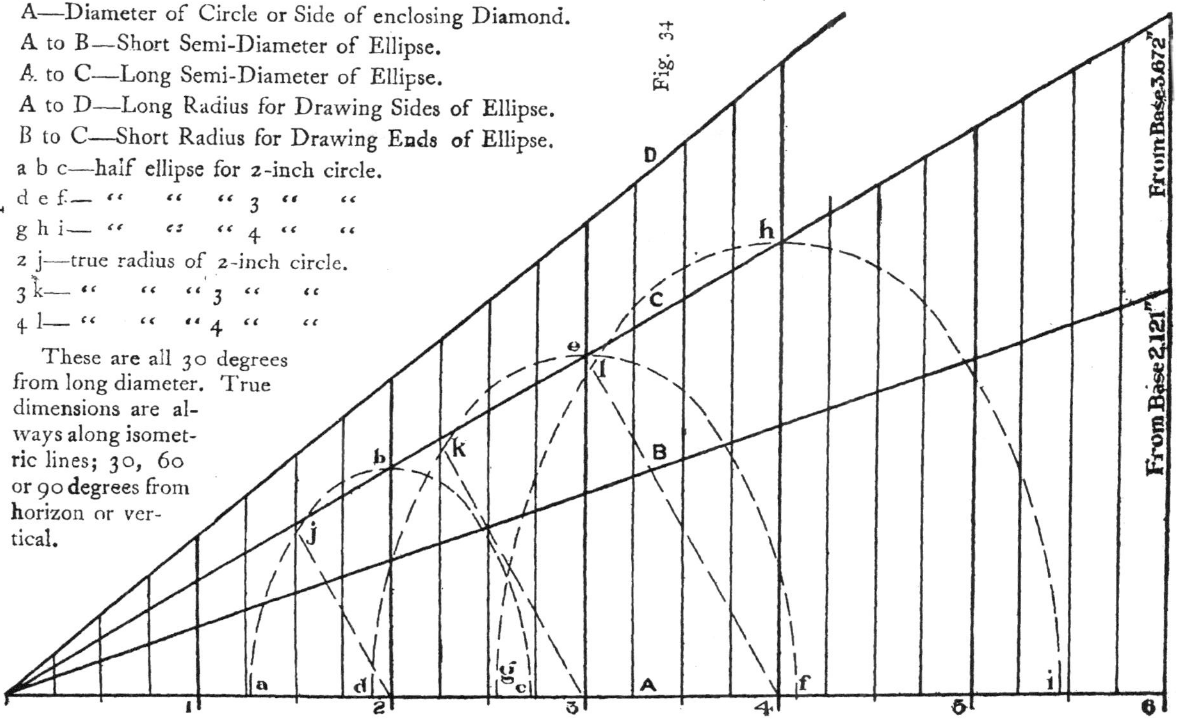

Fig. 34

ruling is very lightly printed so as not to be in the way, but the diamonds are there and they will help you draw ellipses easily and rapidly, either free-hand or with instruments.

One more elementary lesson and we will get down to real drawing of things we see every day. To still further show how the ellipse affects things in isometric drawing and the value of these ellipses to give us the proper dimensions of other parts, we show a box in Fig. 35 with covers both top and bottom. This practically explains itself, as it is easily seen that the length of the sides of the cover is always determined by the ellipse. In the position indicated by the long diameter the cover appears longer than it really is, as the correct dimension is along the isometric lines shown by the fully drawn cover or the sides of the box. This will be found useful in many cases and can be referred to for guidance at any time. This has a close relation to the drawing of hexagon and other nuts, and it may be well to get these straight before we start to draw any machine details.

Fig. 36 shows a cube with regular hex nuts on the two sides and a hex flanged nut on top. The flange is simply two ellipses drawn one above the other and the hexagon is located in its proper place above. This is one of the puzzling things at first. This will be cleared up a little later and the ruling of the paper makes it easy after we understand what is wanted.

Fig. 37 shows the corner of a cube and the ellipses laid out on the three surfaces. Hexagons or squares are laid out practically as in any drawing except that we must only consider the center lines and remember that these are always at the isometric angle. Taking the ellipse on top we sketch in the diameters *A, B* and *C, D*. Then for con-

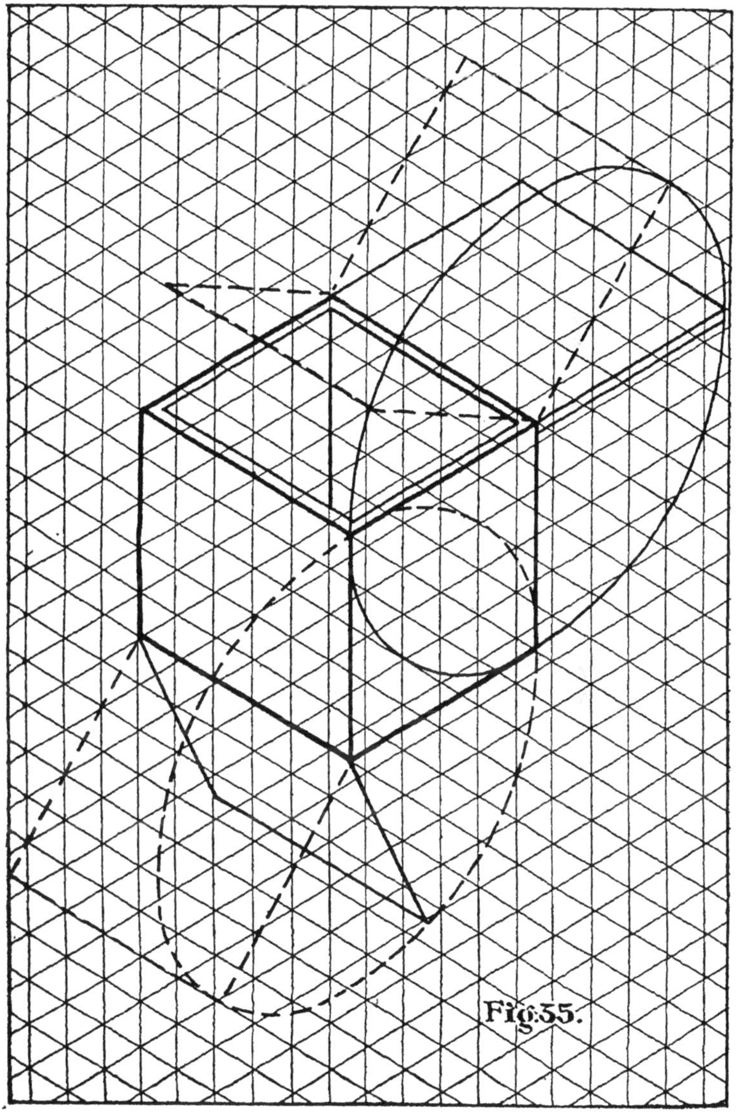

Fig. 55.

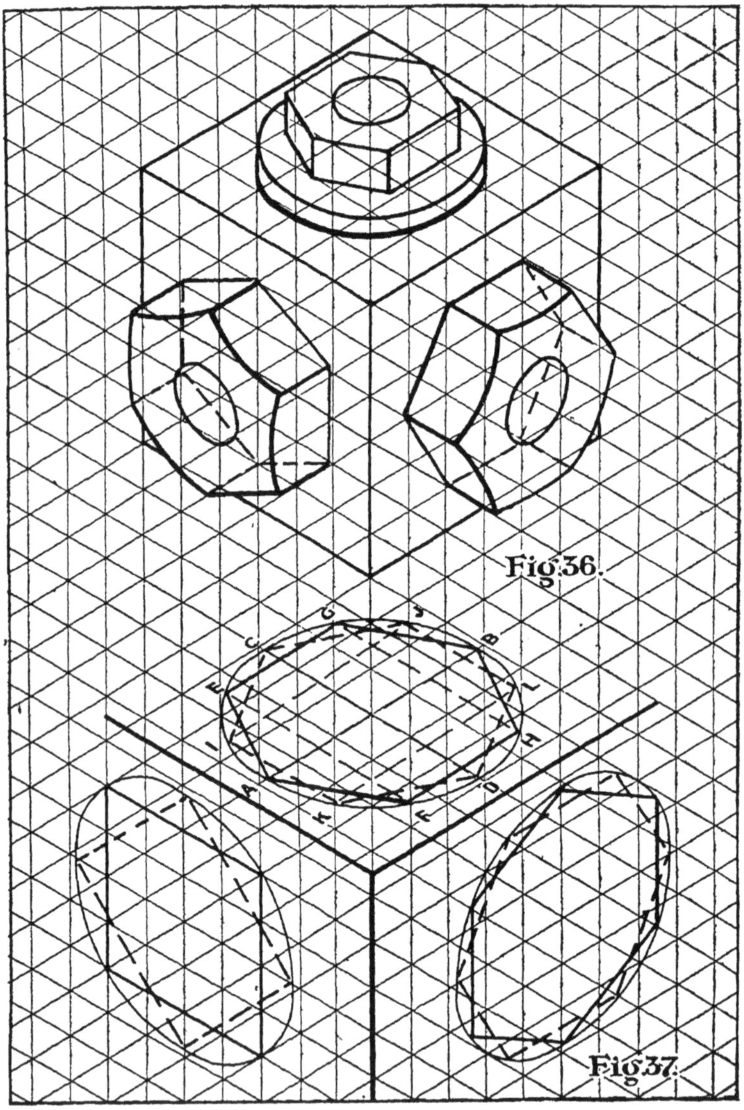

Fig. 36.

Fig. 37.

venience sketch in the lines *E F*, *G H*, *I J*, and *K L*. Where these cut the ellipse give the points for the two hexagons thrown into the right perspective. Connect *A E G B H F* and *A* as shown in solid lines and you have the hexagon whose true long diameter "across the corners" is *A B*. Connecting *I C J L D K* and *I* as with the dotted lines gives the hexagon turned so as to bring the true long diameter *C D*. After we get the idea it will be unnecessary to draw the construction lines at all, but we can just follow the ruling of the paper and locate all the points in less time than it takes to tell of it here. For completing the nut we simply erect vertical lines from the six points to the required thickness of the nut and connect them at the top and you have your nut complete. If it is a nut with a rounded edge it will appear as shown on sides of Fig. 36.

The same method will lay out the hexagon nut for you in any position bearing in mind the same proportions as you always use in laying out a hexagon nut, i. e. with the flat side towards you; this flat side is as long as the other two which you see. A square nut is shown on the left side of Fig. 37, and with this in mind there will be no need to spend further time on this detail. If it were not for the ruled lines on this paper it would be necessary to spend much more time on it and might be well to give a table of proportions for the different styles, but the ruling makes all this superfluous and as will be seen makes it easy for anything to be laid out after one understands the principles of isometric as laid down by Mr. Richards.

In Fig. 38 we try our hand at a sketch of what might be called a square crank. Taking *A B* as the diameter across the corners of the shaft we sketch up to *C D* where the crank starts off to the right. Complete the diamonds

Fig. 38.

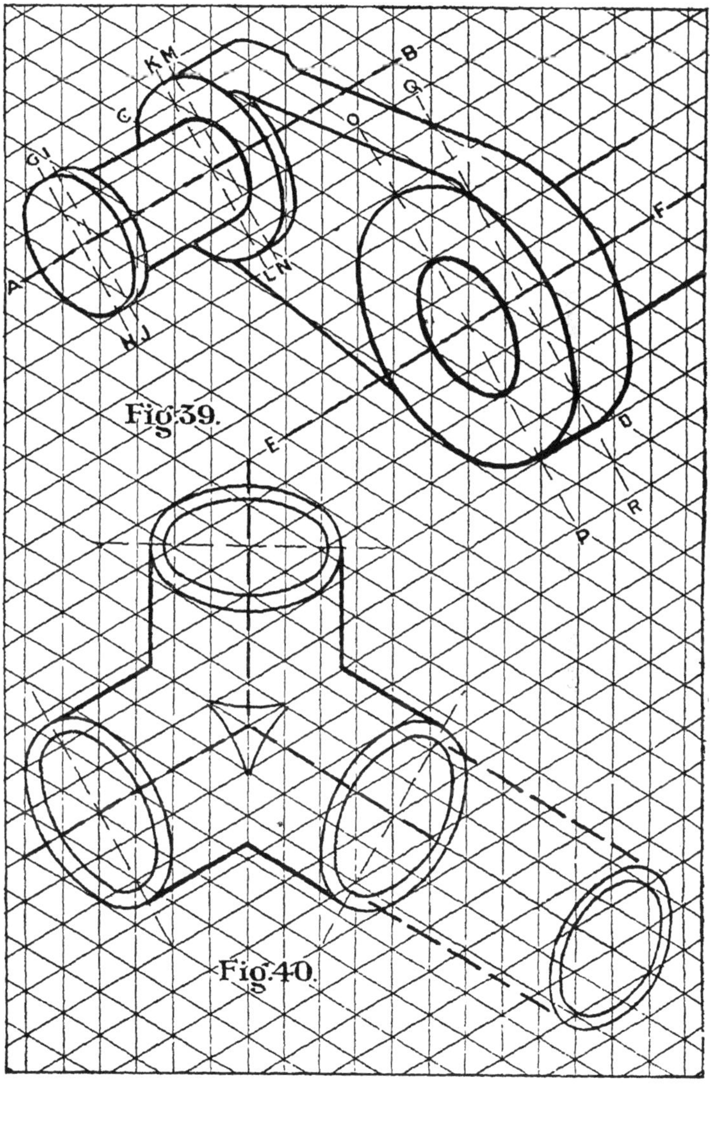

Fig. 39.

Fig. 40.

either in imagination or by dotted lines as a guide. Laying off the offset of the crank *X* we spot out the points *E F* and sketch up to the other side of the crank at *G H* at a distance *Z* from the inside face of the lower part of the crank. Setting out points *I J* we have the continuation of the main shaft and can then look after the crank cheeks. These can be made perfectly plain or can be offset as shown, but there will be no difficulty in laying them off right and quickly if we get started right with a few guide points as shown, and after one gets the run of this paper it can be done without a moment's hesitation. To make this regular crank shaft all that is necessary is to sketch ellipses over the diamonds and the square shafts become round. This is given simply as an example in laying out the different points so as to get things just right.

Going to Fig. 39 we have a genuine crank shaft of the engine variety. The first thing to do is to lay the center lines for the crankpins, the line from this to the center of the shaft and the center line of the shaft. These are *A B, C D* and *E F.* The center line of the ellipse, *G H,* locates the outer end of the collar on the crank and *I J* for the inside of this collar. The distance between these lines is the thickness of the collar. From *I* to *K* is the length of the crank pin, and from *K* to *M* the height the hub extends above the web of the crank. The ellipse around *K L* gives the face of crank hub or boss, and around *M N* (or the portion which would show) the corner in the crank between web and hub. Center lines, *O P* and *K L* both cross their center lines along the line *C D,* and the distance between *A* and *E* is throw of the crank. Distance between *O* and *Q* is the thickness of the main hub or crank, and the ellipses complete the sketch. A little practice will enable anyone to make anything of this

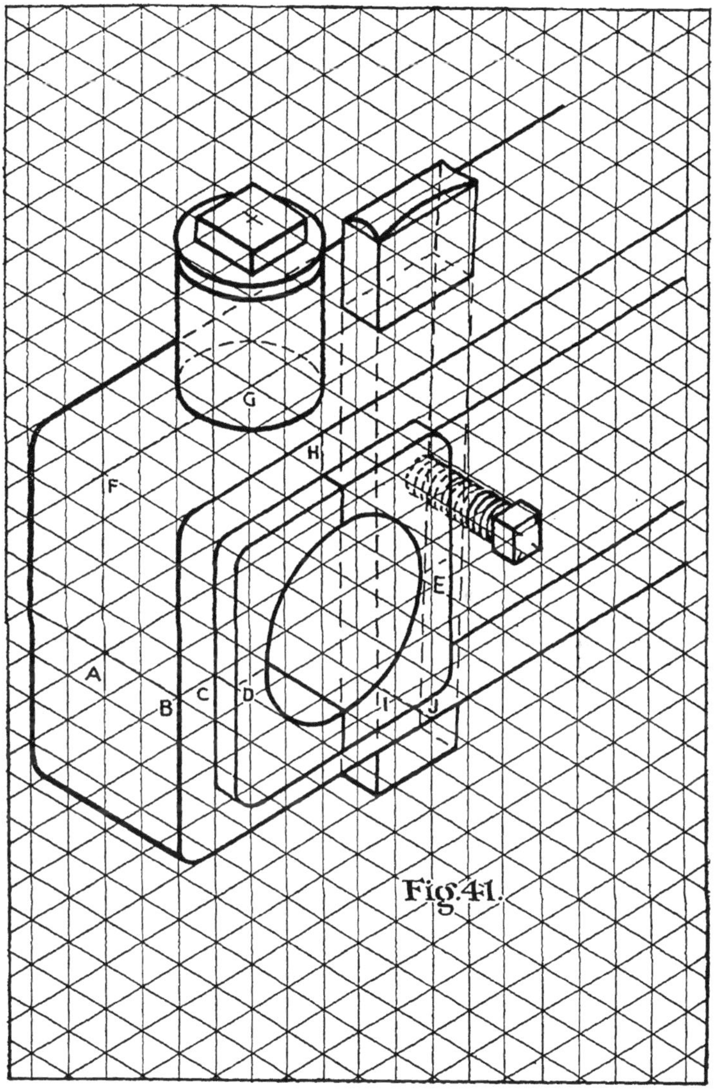

Fig. 41.

kind that may come up in regular work, in a very short time, and there isn't a man in the shop who cannot see at a glance just what is wanted much better than with the regular three view sketch.

Fig. 40 shows a three way elbow with a piece of pipe in one end and illustrates the position of the ellipses for all three directions as well as any simple everyday object that we can find.

Fig. 41 shows the business end of a connecting rod of a locomotive or other engine. The only things that will bother at all in doing this is the offset of the rod brasses which projects outside the strap and the set screw and key. The oil cup is easy when we remember that the ellipse is horizontal in all cases of this kind. As a guide in seeing just how this was laid out we have put a few construction lines which, however, were not used in making the drawing as they are not necessary after one has a little experience with the use of the paper. The dotted lines showing the hidden parts of the key and oil cup were put in to make clear just how the different points were obtained and to show that they are located correctly.

Figs. 42 and 43 show wrenches for shop use. The monkey wrench is of a knife handle variety and not drawn to any special scale but to show that it can be readily handled in this way. The dotted lines at the head clear up any confusion as to proper location of lines there, while method of locating the rivet in the handle is shown by the dotted lines there. The distance from the center line to the center line of the rivet is half the thickness of the handle.

In the special wrench below we have a square opening socket wrench with a piece of pipe through to four hole

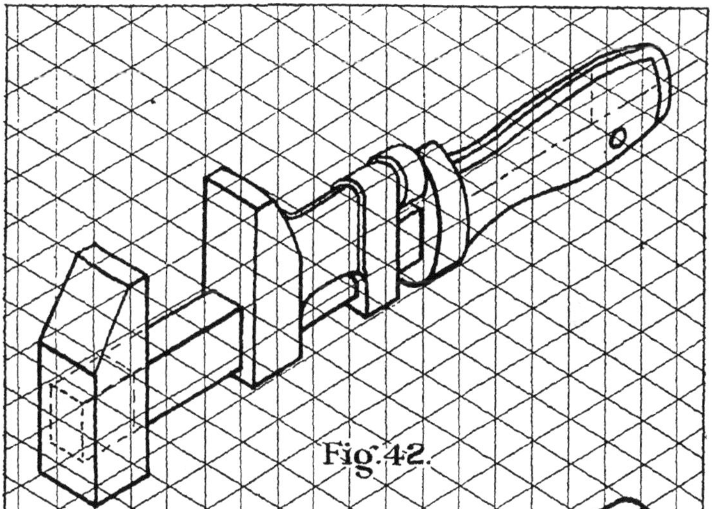

Fig. 42.

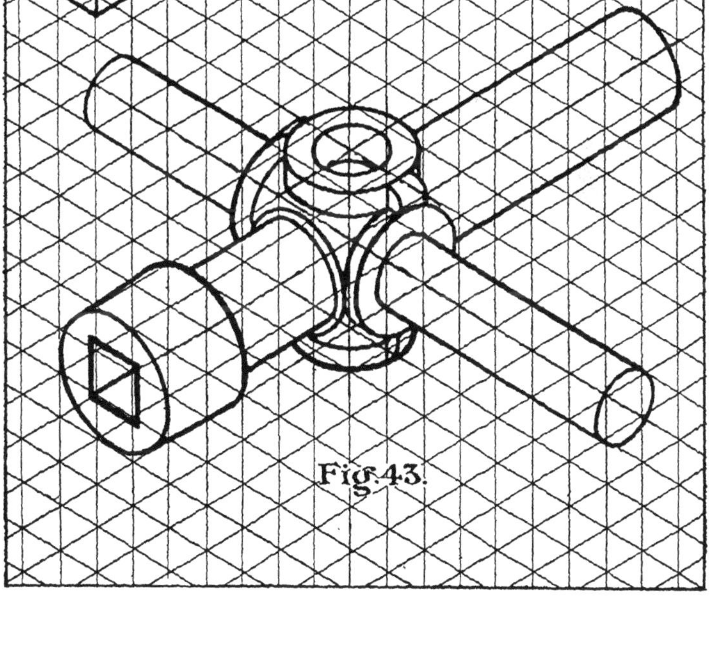

Fig. 43.

opening in the handle. This would be rather a difficult thing to show so a workman who was not familiar with drawing could readily understand if we used the regular three view projection method, but there is not a man who calls himself a machinist who would hesitate a minute on such a sketch as this. You can dimension every part that you need to and give it to the mechanic without any fear of his going astray on the job. In fact this is the really useful field for this method of drawing and the paper here described. It gives a means of making rapid and accurate sketches freehand that can be given to any workman with a knowledge that he can understand it. If finished drawings are wanted they can be made as we have seen, but one of the most useful fields is the drawing room sketch. This can either be copied in a press or blue printed as desired.

The next sheet shows a number of rough free hand sketches. Needless to say they were not made by an expert and can be equalled by any one in the drawing room and vastly improved upon by all who are at all expert in free hand work.

The regulation two-jawed chuck in Fig. 44, the core-box in 45, the end of an *I* beam with a few details of connections, all explain themselves. Just try to sketch out the pipe connections shown in Fig. 47 by the three view process and see if you could understand it yourself if it was sprung on you suddenly. Would any plumber or steam fitter have any trouble in knowing just what you wanted from the sketch shown? And this could have been elaborated almost without limit and yet be clear because you can see everything and put down any dimension you want to.

Fig. 48 might be a sample block of joints for a

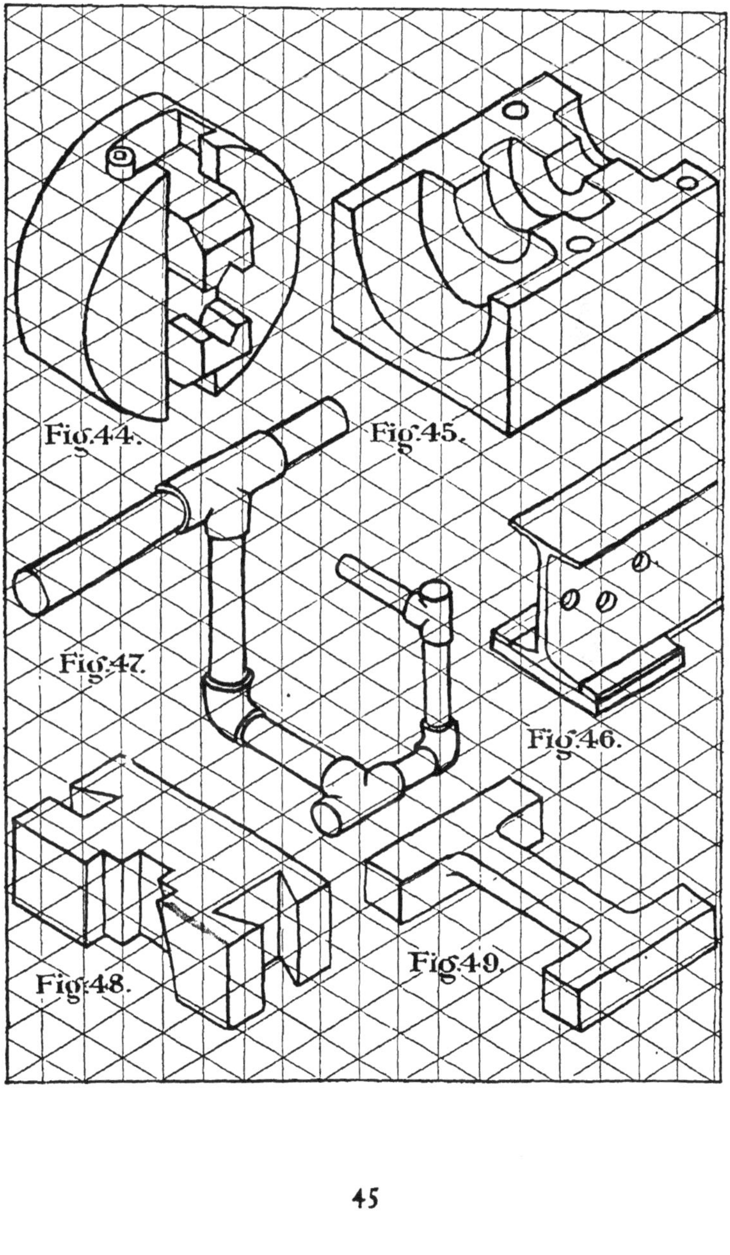
Fig.44.
Fig.45.
Fig.47.
Fig.46.
Fig.48.
Fig.49.

manual training school or a key piece to a puzzle, but whatever it was there would be no trouble in any carpenter or machinist making a piece like it out of either wood or iron if the dimensions were given, as might easily be done.

A blacksmith would welcome a sketch like Fig. 49, for he could see exactly what he wanted. These might be multiplied many times, but these are sufficient to show what can be done with a very little practice.

The three following figures show details of different kinds of machine construction which do not need to be described in detail, and the ruling lines are given to show how easily the work may be done. In fact the work can, for the most part be done without even the aid of a rule. Simply call each space between lines an eighth of an inch, if it is small work, an inch or a yard, so long as it is always the same and you can lay out anything you want with only a pencil and a rubber. The latter not to hide mistakes but to erase construction lines which often make the work easier.

Fig. 53 shows the assembly and details of a head-stock. There is no need of detailed explanation as any mechanic can see just what it is.

Taking Fig. 54 as an example and we find a piece of structural steel construction work which is surely clear for any one. The shapes used and the methods of joining are made very clear in a sketch like this and men in the field would have no hesitation in following the drawings without thought of a mistake.

Beginning at the bottom plate we outline it, go up a half space for its thickness and another half space for the lower angle used as the fastening. Locate where the *Z* bars will come by means of the ruling and lay out their shape at the top of the sketch with the plate between them

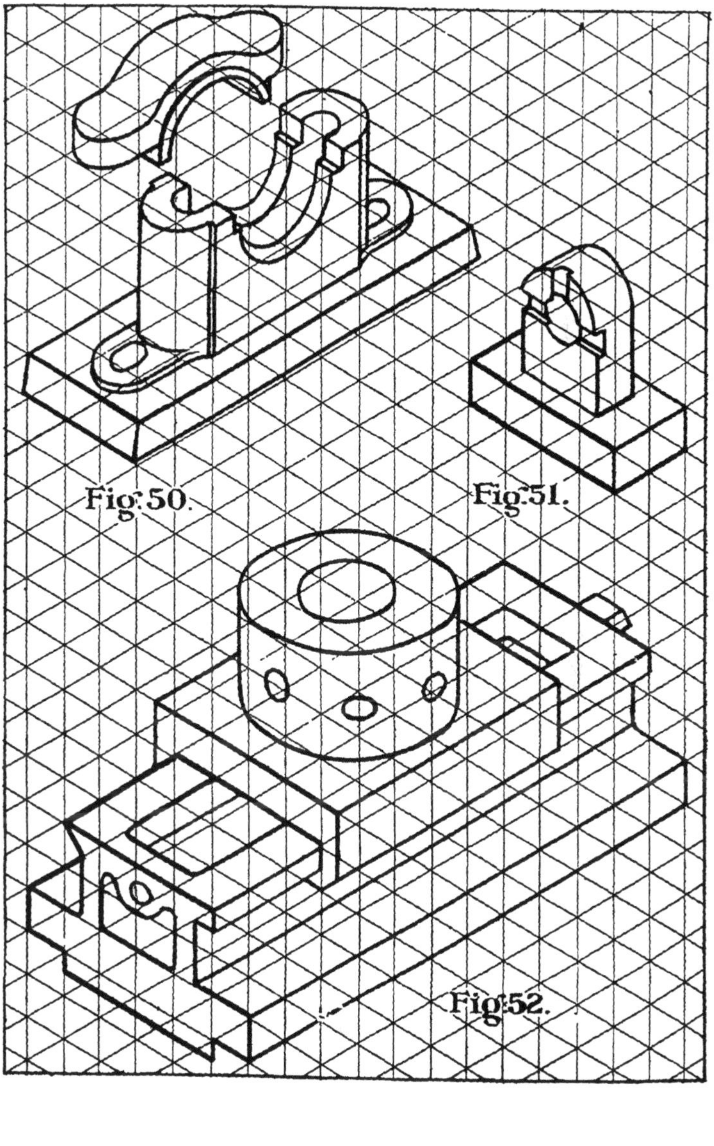

Fig: 50. Fig: 51. Fig: 52.

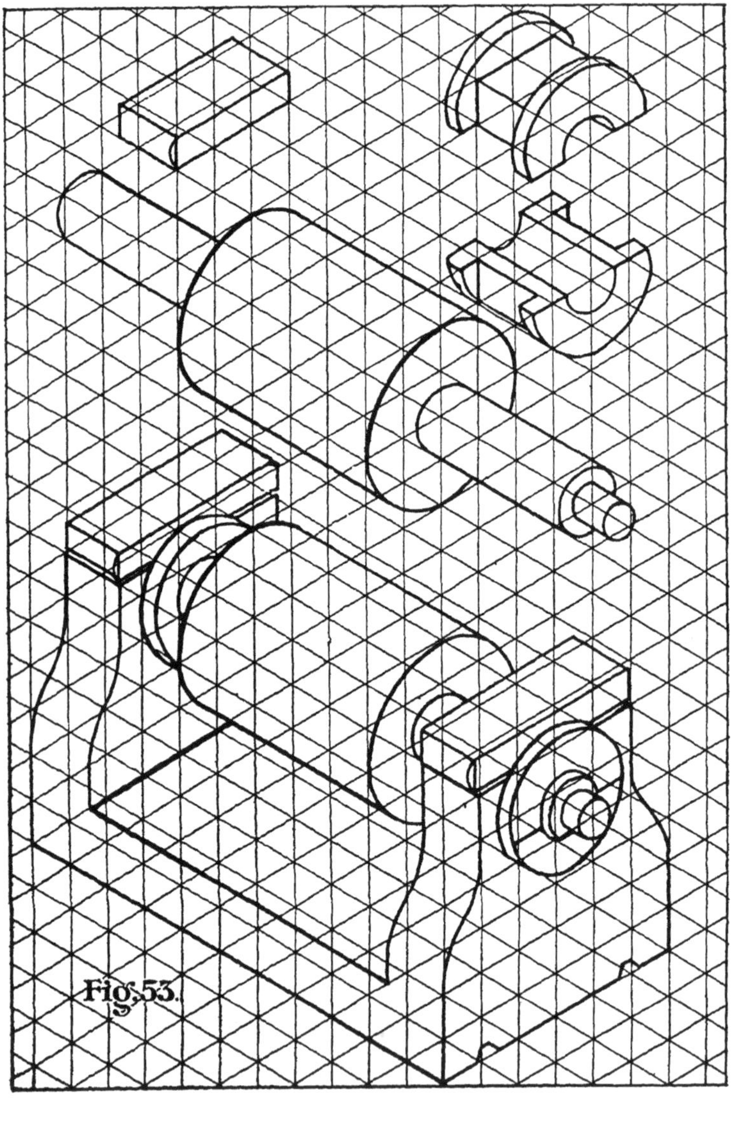

Fig. 53.

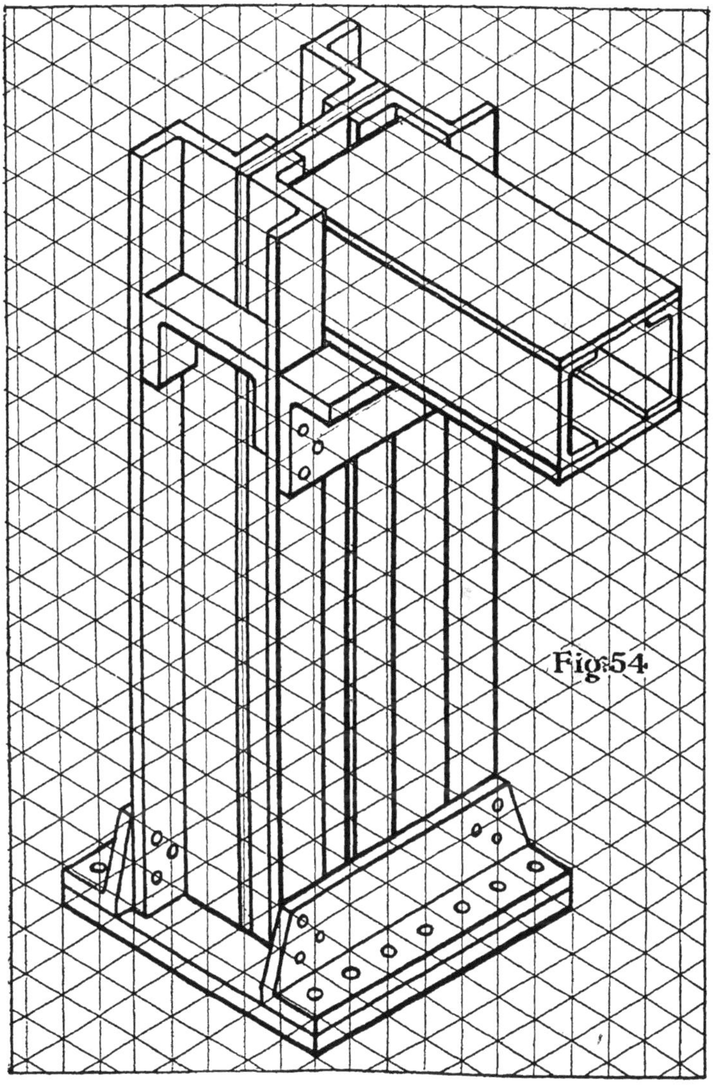

Fig. 54

to form a web as shown. Then the box beam, made up of two channels and two plates is located in between the *Z* bars and the supporting angle put underneath. The ruling of the paper takes away all the necessity for *T* squares or angles and you can lay off distance or offsets by simply following the lines in the right direction. This will sometimes be up and across to get the right offset, but you can always tell when you have it right or wrong, as it shows at a glance when you get a little used to it. Everything that is square or rectangular will come out exactly in diamonds every time and if it doesn't you will find some point which isn't quite exactly right.

Architectural details can also be very nicely handled in this way, even complete drawings as will be seen from some of the illustrations given further along in the book to show what has been done with it.

Fig. 55 shows the corner of a building with the cornice and its details all drawn free hand. These ellipses are oposite from any others we have used as their lines of perspective run down in each case. If you ever have any doubts as to the position, sketch one in lightly and you can tell at a glance whether it is right or wrong.

Fig. 56 is from a set of concrete steps and railings from recent practice in railroad station work, and simply shows another aplication of this method of projection.

The illustrations which follow are all drawn in this perspective, some with this paper and some without it, although the ruling is omitted to show how they look without it and also how much more difficult it is to draw with the specially ruled paper. As each illustration carries its own explanation there is no need of further description in the text. With these examples, which include nearly all branches of mechanics, it should not be difficult to find

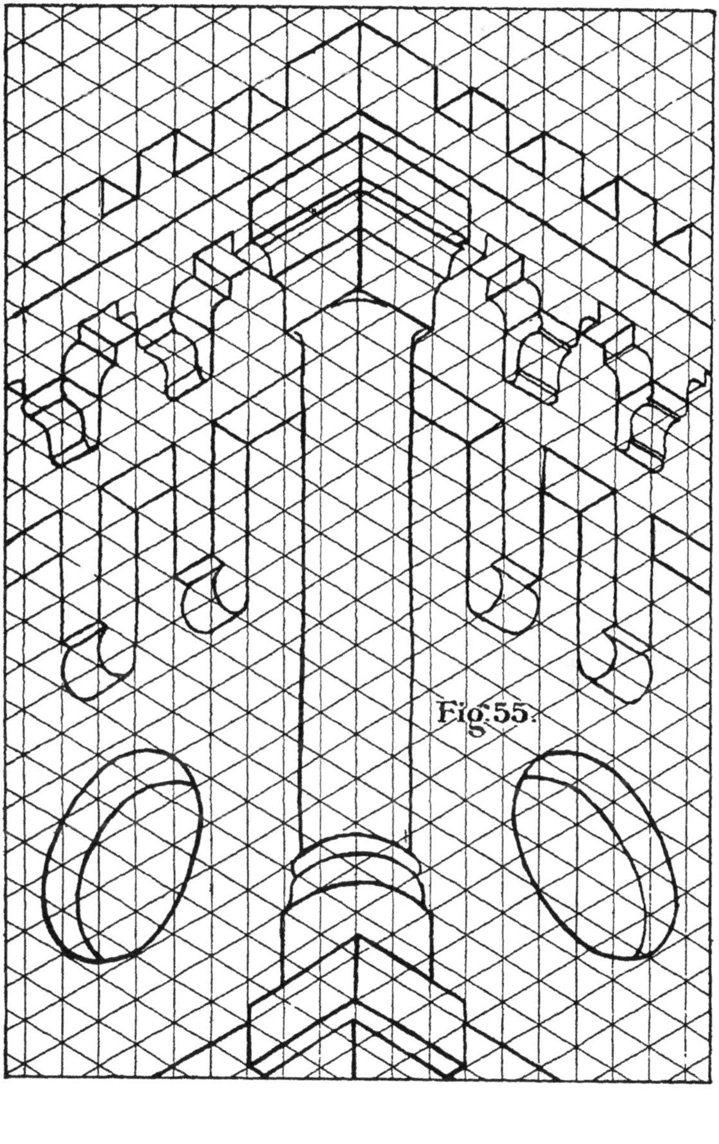

Fig. 55.

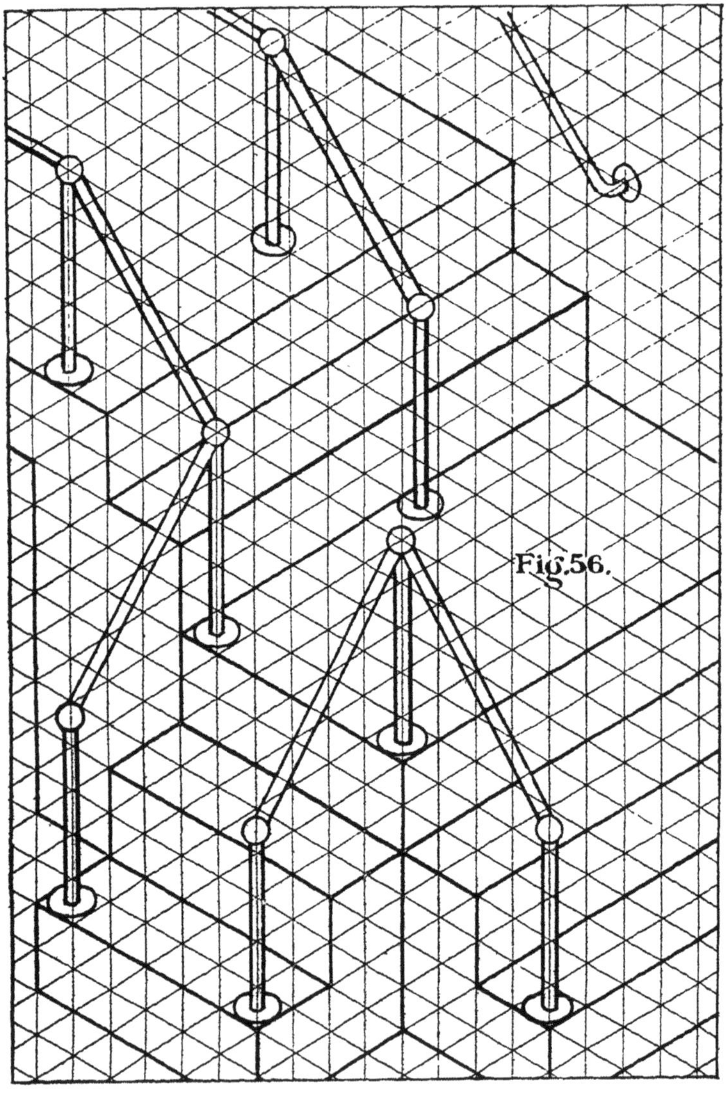

Fig. 56.

illustrations of almost anything you wish to make so as to see exactly how to go to work if you have any hesitation in the matter. It's largely a matter of common sense and a little patience, and if you follow the lines, counting the spaces to get your various offsets, there will be no difficulty in handling even complicated drawings in this way.

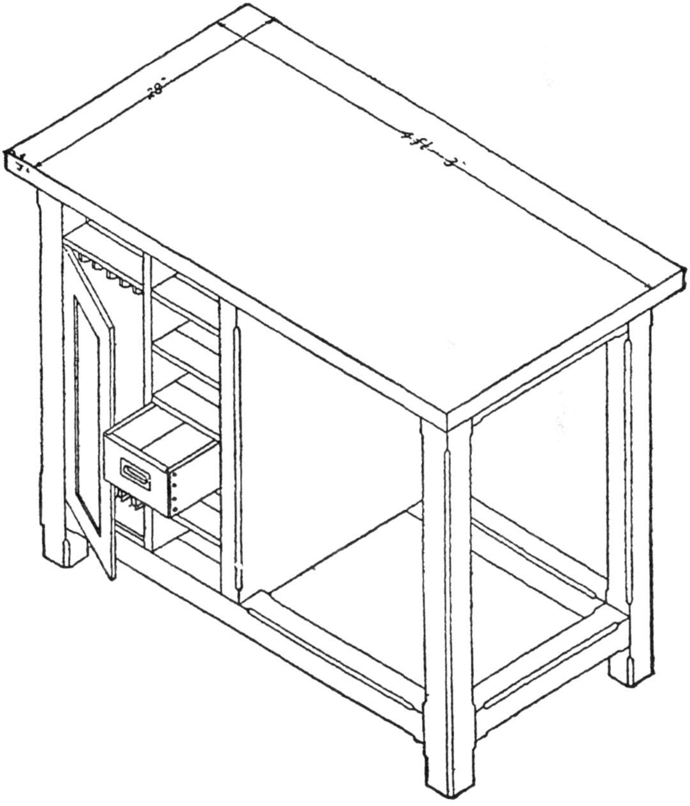

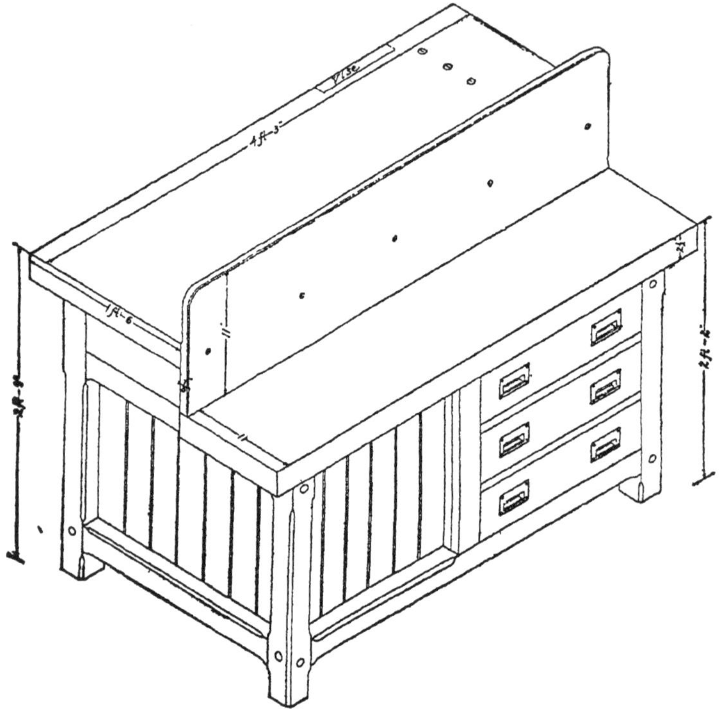
Vise
4 ft-3"
1 ft-6
2 ft-8"

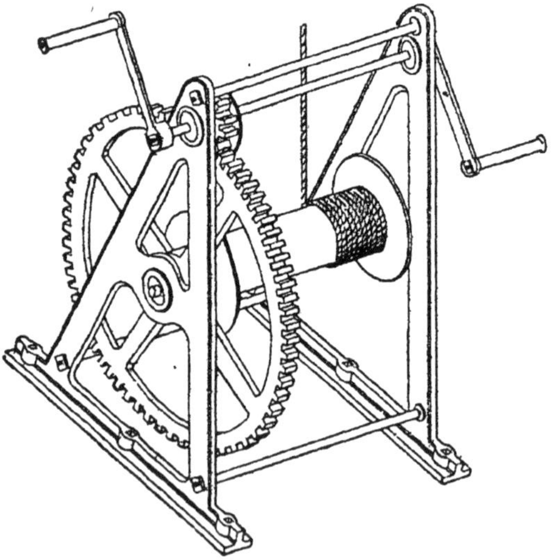

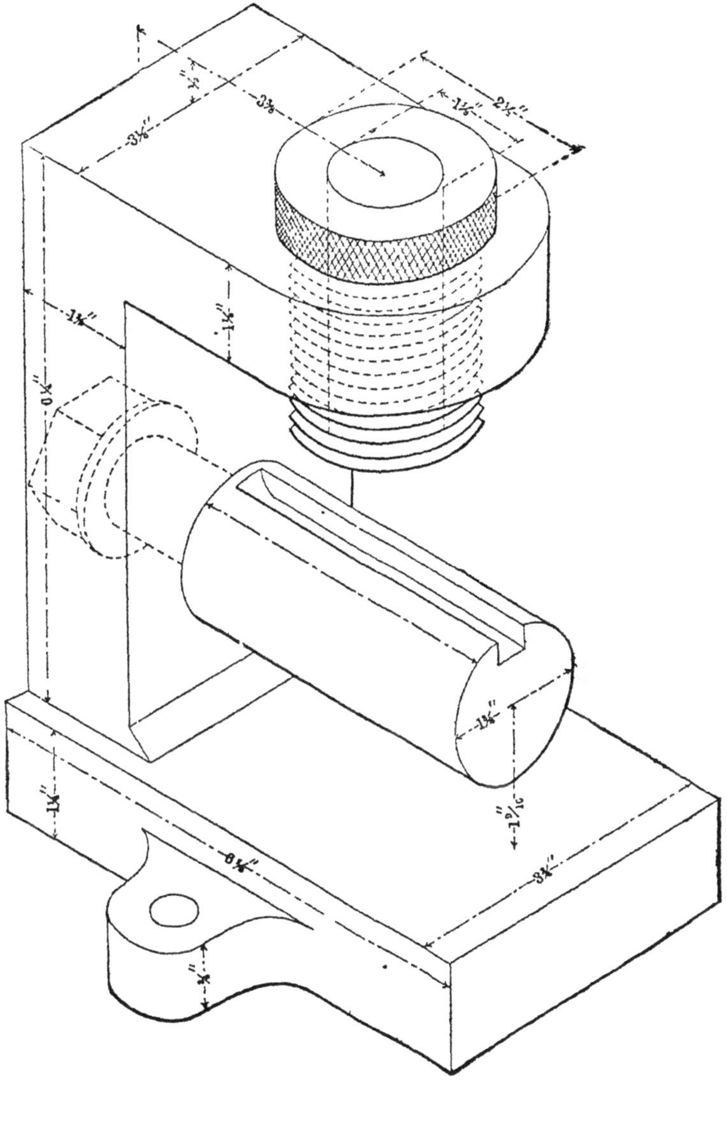

The Use of Isometric Paper.

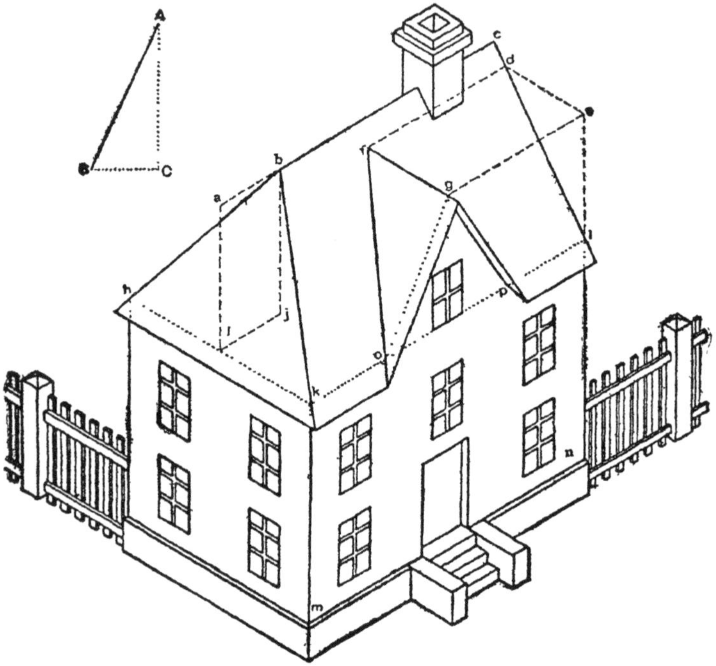

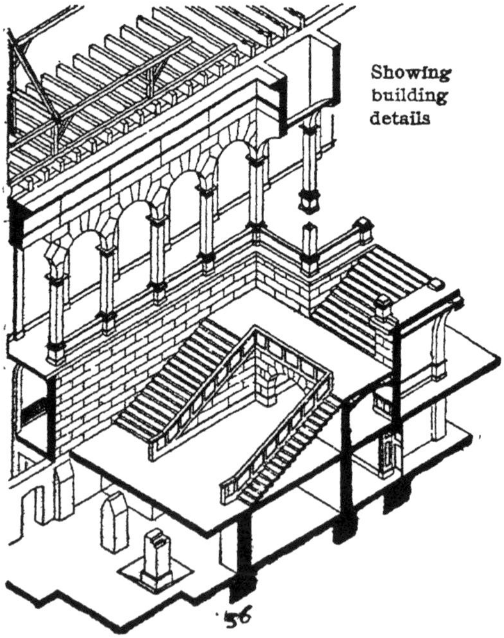

Showing
building
details

BLACKALL. New York Air-Brake Catechism.
A complete treatise on the New York Air-Brake and Air Signalling Apparatus, giving a detailed description of all the parts, their operation, troubles, and the methods of locating and remedying the same. 200 pages, fully illustrated. $1.25.

BOOTH AND KERSHAW. Smoke Prevention and Fuel Economy.
A complete treatise for all interested in smoke prevention and combustion. Mr. Booth was formerly of the Manchester Steam Users' Association, is a well known writer upon technical subjects, and his associate is a practical chemist. The co-operation of the authors has been very successful and the principles of fuel combustion and the means by which the practice is conducted are briefly but adequately discussed. The physical and chemical conditions involved are given due consideration and there are descriptions of American and English appliances, mechanical stokers, etc. 194 pages. 75 illustrations. Price, $2.50.

Booth Steam Pipes, Their Design and Construction.
A treatise on the principles of steam conveyance and means and materials employed in practice, to secure economy, efficiency and safety. A book which should be in the hands of every steam fitter and user of steam; contains 177 pages on the use of steam for power and other purposes. Price, $2.00.

BUCHETTI. Engine Tests and Boiler Efficiencies.
This work fully describes and illustrates the method of testing the power of steam engines, turbine and explosive motors. The properties of steam and the evaporative power of fuels. Combustion of fuel and chimney draft; with formulas explained and practically computed. 255 pages; 179 illustrations. $3.00.

COCKIN. Practical Coal Mining.
An important work, complete with practical details, which will intuitively impart to the reader not only a general knowledge of the principles of coal mining, but also considerable insight into allied subjects, including chemistry, mechanics, steam and steam engines and electricity. 428 pages, 213 illustrations. Price, $2.50.

FOWLER. Locomotive Breakdowns and Their Remedies.
This work treats in full all kinds of accidents that are likely to happen to locomotive engines while on the road. The various parts of the locomotives are discussed, and every accident that can possibly happen, with the remedy to be applied, is given. 250 pages, fully illustrated. $1.50.

FOWLER. Boiler Room Chart.
An educational chart showing in isometric perspective the mechanisms belonging in a modern boiler room. The chart serves as a dictionary of the boiler-room, the names of more than two hundred parts being referred to. 25 cents.

GRIMSHAW. Saw Filing and Management of Saws.
A practical handbook on filing, gumming, swaging, hammering, and the brazing of band saws, the speed, work and power to run circular saws, etc. Fully illustrated. Cloth, $1.00.

GRIMSHAW. "Shop Kinks."
This book is entirely different from any other on machine-shop practice. It is not descriptive of universal or common shop usage, but shows special ways of doing work better, more cheaply and more rapidly than usual, as done in fifty or more leading shops in Europe and America. Some of its over 500 items and 222 illustrations are contributed directly for its pages by

eminent constructors; the rest have been gathered by the author in his thirty years' travel and experience. Fourth edition. Nearly 400 pages. Cloth, $2.50.

GRIMSHAW. Engine Runner's Catechism.

The questions asked throughout the catechism are plain and to the point, and the answers are couched in such simple and homely language as to be readily understood by the veriest dullard. All the instructions given are complete and up to date; and they are written in a popular style, without any technicalities or mathematical formulae. The work is of a handy size for the pocket, clearly and well printed, nicely bound, and profusely illustrated. To young engineers this catechism will be of great value, especially to those who may be preparing to go forward to be examined for certificates of competency; and to engineers generally it will be of no little service, as they will find in this volume more really practical and useful information than is to be found anywhere else within a like compass. Tells how to erect, adjust and run the principal steam engines in the United States. Sixth edition. 336 pages. Fully illustrated. Cloth, $2.00.

GRIMSHAW. Steam Engine Catechism.

This unique volume is not only a catechism on the question and answer principle; but it contains formulas and worked out answers for all the steam problems that appertain to the operation and management of the steam engine. Illustrations of various valves and valve gear with their principles of operation are given. Thirty-four tables that are indispensable to every engineer and fireman that wishes to be progressive and is ambitious to become master of his calling are within its pages. It is a most valuable instructor in the service of steam engineering. Leading engineers have recommended it as a valuable educator for the beginner as well as a reference book for the engineer. It is thoroughly indexed for every detail. Contains nearly 1,000 questions with their answers, mainly intended for young engineers and for examination questions. Fourteenth edition. 413 pages. Fully illustrated. $2.00.

GRIMSHAW. Locomotive Catechism.

This is a veritable encyclopaedia of the locomotive, entirely free from mathematics, and thoroughly up-to-da e. It contains 4,000 questions with their answers. Twent -seventh edition, greatly enlarged. Nearly 900 pages, over 437 illustrations, and 3 large folding plates. Cloth, $2.50.

HISCOX. Gas, Gasoline and Oil Engines.

Every user of a gas engine needs this book. Simple, instructive and the only complete work on this important subject. Tells all about the running and management of gas engines. Full of general information about the new and popular motive power, its economy and ease of management. Also chapters on horseless vehicles, electric lighting, marine propulsion, etc. 412 pages. Illustrated with 312 handsome engravings. Twelfth edition, revised and enlarged. $2.50.

HISCOX. Compressed Air in All Its Applications.

The most comprehensive book on the subject, showing different types of compressors as well as tools driven by compressed air. Contains a vast amount of general information. 545 illustrations. 820 pages. Price, cloth, $5.00; half morocco, $6.50.

HISCOX. Horseless Vehicles, Automobiles and Motor Cycles.

The make-up and management of automobile vehicles of all kinds are treated. A complete list of the automobile and motor manufacturers, with their addresses, as well as a list of patents issued since 1856 on the automobile industry, are included. Nineteen chapters. Large 8vo. 316 illustrations. 460 pages. Cloth, $3.00.

HISCOX. Mechanical Movements, Powers and Devices.

A work on illustrated mechanics, mechanical movements, powers and devices, covering nearly the whole range of the practical and inventive field, for the use of mechanics, inventors, engineers, draughtsmen and all others interested in any way in mechanics. Large 8vo. Over 400 pages. 1,800 specially made illustrations, with descriptive text. Tenth edition. $3.00.

HISCOX. Mechanical Appliances, Mechanical Movements and Novelties of Construction.

A complete and supplementary volume to the author's work, "Mechanical Movements, Powers and Devices." Contains 1,000 mechanical details and complex combinations in mechanical construction, which are fully described and illustrated 400 pages. $3 00.

Inventor's Manual; How to Make a Patent Pay.

This is a book designed as a guide to inventors in perfecting their inventions, taking out their patents and disposing of them. 119 pages. Cloth. $1.00.

KRAUSS. Linear Perspective Self-Taught.

The underlying principle by which objects may be correctly represented in perspective is clearly set forth in this book; everything relating to the subject is shown in suitable diagrams, accompanied by full explanations in the text. Price, $2.50.

LE VAN. Safety Valves; Their History, Invention and Calculation.

Illustrated by 69 engravings. 151 pages. $1.50.

LEWIS AND BRAME. Laboratory Note Book for Chemical Students.

This volume contains instructions for the laboratory preparation of gases, the examination of simple salts, the reaction of metals and acid radicals, and includes a number of experiments in quantitative analysis. Price, $1.50.

MATHOT. Gas Engines and Producer Gas Plants.

A Guide for the Gas Engine Designer, User and Engineer in the construction, selection, purchase, installation, operation and maintenance of gas engines.

Recognizing the need of a volume that would assist the gas engine user in understanding thoroughly the motor upon which he depends for power, the author has discussed his subject without the help of any mathematics and without elaborate theoretical explanations. Every part of the gas engine is described in detail, tersely, clearly, with a thorough understanding of the requirements of the mechanic. Helpful suggestions as to the purchase of an engine, its installation, care and operation, form a most valuable feature of the work. Fully illustrated. $2.50.

PARSELL AND WEED. Gas Engine Construction.

A practical treatise describing the theory and principles of the action of gas engines of various types, and the design and construction of a half-horse power gas engine, with illustrations of the work in actual progress, together with dimensioned working drawings, giving clearly the sizes of the various details. Second edition, revised and enlarged. Large 8vo. 300 pages. $2.50.

REAGAN, JR. Electrical Engineers' and Students' Chart and Hand Book of the Brush Arc Light System.

Illustrated. Bound. in cloth, with celluloid chart in pocket. $1.00.

SLOANE. Electricity Simplified.

This work is the simplest ever published on the subject of electricity, and does something not hitherto accomplished. Electricity is in many respects unexplained by the scientist; to the ordinary man it is all a mystery. The object of "Electricity Simplified" is to make the subject as plain as possible, and to show what the modern conception of electricity is; to show how two plates of different metals immersed in acid can send a message around the globe; to explain how a bundle of copper wire rotated by a steam engine can be the agent in lighting our streets; to tell what the volt, ohm, and ampere are, and what high and low tension mean; and to answer the questions that perpetually arise in the mind in this age of electricity. The theories of contact action, of lines of force, magnetic permeability, correlation and conservation of energies, and the most modern aspects of the science are given so that the work is a true presentation of the most advanced views of science upon the great development of the day. By illustration of original design and scope, and by mechanical analogies, the subject is made exceedingly simple. 172 pages. Illustrated. Twelfth edition. $1.00.

SLOANE. How to Become a Successful Electrician.

It is the ambition of thousands of young and old to become electrical engineers. Not everyone is prepared to spend several thousand dollars upon a college course, even if the three or four years requisite are at their disposal. It is possible to become an electrical engineer without this sacrifice, and this work is designed to tell "How to Become a Successful Electrician" without the outlay usually spent in acquiring the profession. Twelfth edition. 202 pages. Illustrated. Cloth. $1.00.

SLOANE. Arithmetic of Electricity.

A practical treatise on electrical calculations of all kinds, reduced to a series of rules, all of the simplest forms, and involving only ordinary arithmetic; each rule illustrated by one or more practical problems, with detailed solution of each one. Sixteenth edition. Illustrated. 162. pages. Cloth. $1.00.

SLOANE. Electrician's Handy Book.

This book supplies a distinct want in the realm of electrical literature. It is designed to cover the field of practical engineering, yet to include nothing unnecessary for the every day worker in electricity to know. Its pages are not encumbered with any useless theory—everything in it is to the point and can be readily understood by the non-technical man, and at the same time the educated engineer will receive great benefit from its perusal. It is a modern book of reference, a compendium of useful data. It gives the clue to the operation of electrical systems of to-day leaving out the old and useless matter which has encumbered many text books, yet not omitting hints from the past which have a meaning in the present. The latest and best authority on all branches of applied electricity. Pocketbook size. Handsomely bound in leather, with title and edges in gold. 800 pages. 500 illustrations. Price, $3.50.

SLOANE. Rubber Hand Stamps and the Manipulation of Rubber.

A practical treatise on the manufacture of all kinds of rubber artices. 146 pages. Second edition. Cloth. $1.00.

SLOANE. Electric Toy Making, Dynamo Building, and Electric Motor Construction.

This work treats of the making at home of electrical toys, electrical apparatus, motors, dynamos, and instruments in general, and is designed to bring within the reach of young and old the manufacture of genuine and useful electrical appliances. Fifteenth edition. Fully illustrated. 183 pages. Cloth. $1.00.

SLOANE. Liquid Air and the Liquefaction of Gases.

Containing the full theory of the subject and giving the entire history of liquefaction of gases from the earliest times to the present. 365 pages, with many illustrations. Second edition. $2.50.

SLOANE. Standard Electrical Dictionary.

A practical handbook of reference, containing definitions of about 5,000 distinct words, terms and phrases. An entirely new edition, brought up to date and greatly enlarged. Complete, concise, convenient. 682 pages. 393 illustrations. $3.00.

USHER. The Modern Machinist.

A practical treatise embracing the most approved methods of modern machine-shop practice, and the applications of recent improved appliances, tools and devices for facilitating, duplicating and expediting the construction of machines and their parts. A new book from cover to cover. Fifth edition. 257 engravings. 322 pages. Cloth. $2.50.

VAN DERVOORT. American Lathe Practice.

A new book from cover to cover. It is strictly up-to-date in its descriptions and illustrations, which represent the very latest practice in lathe and boring-mill operations as well as the construction of and latest developments in the manufacture of these important classes of machine tools. A large amount of space is devoted to the turret lathe, its modifications and importance as a manufacturing tool. 320 pages. 200 illustrations. $2.00.

VAN DERVOORT. Modern Machine Shop Tools; Their Construction, Op-ration and Manipulation.

This is a book of reference that will be found convenient in every machine shop. Suppose it is desired to know how to cut bevel gears, to calculate milling machine spirals or to make countershaft calculations; or to get information about tap drill sizes, the classification of files; change gear calculations; deep hole drilling; turning tapers; testing lathes, etc.; or any one of the numerous questions that a little information might be desired upon occasionally—these pages will be found to contain the satisfactory answer. The book will also prove a boon to students in manual training, as by studying its pages and applying its principles to the school shop, they will be able to acquire a good knowledge of shop practice. The book has numerous tables, and in addition to the chapters strictly on tools are several on fastenings, gearing, belting, shafting, and the treatment of steel. 552 pages and 673 illustrations. $4.00.

WALLIS-TAYLOR. Pocket Book of Refrigeration and Ice Making.

This explains the properties and refrigerating effect of the different fluids in use, the management of refrigerating machinery and the construction and insulation of cold rooms with their required pipe surface for different degrees of cold; freezing mixtures and non-freezing brines, temperatures of cold rooms for all kinds of provisions, cold storage charges for all classes of goods, ice making and storage of ice, data and memoranda for constant reference by refrigerating engineers, with nearly one hundred tables containing valuable references to every fact and condition required in the installment and operation of a refrigerating plant. Price, $1.50.

WOODWORTH. American Tool Making and Interchangeable Manufacturing.

A complete treatise on the Art of American Tool Making and System of Interchangeable Manufacturing as carried on to-day in the United States. In it are described and illustrated all of the different types and classes of small tools, fixtures, devices and special appliances which are in general use in all machine manufacturing and metal working establishments where economy, capacity and interchangeability in the production of machined metal parts are imperative. 500 pages. 600 illustrations. Price, $4.00.

WOODWORTH. Dies, Their Construction and Use for the Modern Working of Sheet Metals.

A practical work on the designing, constructing and use of tools, fixtures and devices, together with the manner in which they should be used in the power press for the cheap and rapid production of sheet metal parts and articles. Comprising fundamental designs and practical points by which sheet metal parts may be produced at the minimum of cost to the maximum of output, together with special reference to the hardening and tempering of press tools and to the classes of work which may be produced to the best advantage by the use of dies in the power press. Fourth edition. 400 pages. 500 illustrations. $3.00.

WOODWORTH. Hardening, Tempering, Annealing and Forging of Steel.

A new book containing special directions for the successful hardening and tempering of all steel tools. Milling cutters, taps, thread dies, reamers, both solid and shell, hollow mills, punches and dies, and all kinds of sheet-metal working tools, shear blades, saws, fine cutlery, and metal-cutting tools of all descriptions, as well as for all implements of steel, both large and small, the simplest and most satisfactory hardening and tempering processes are presented. The uses to which the leading brands of steel may be adapted are concisely presented, and their treatment for working under different conditions explained, as are also the special methods for the hardening and tempering of special brands. 320 pages. 250 illustrations. $2.50.

WRIGHT. Electric Furnaces and Their Industrial Applications.

Contains 285 pages, and 57 illustrations, which are essentially in the nature of sectional diagrams, representing principles of construction. This is a timely and practical treatise on the forms and uses of electric furnaces in modern electro-chemical processes. Price, $3.00.

Made in the USA
Monee, IL
04 November 2026

67388643R10037